# BANANA PHYSICS

# BANANA PHYSICS

## A YOUNG PHYSICIST'S GUIDE TO UNDERSTANDING THE UNIVERSE

ADNAN CONTRACTOR

NEW DEGREE PRESS

BANANA PHYSICS

*A Young Physicist's Guide to Understanding the Universe*

ISBN    978-1-64137-179-7  *Paperback*

978-1-64137-180-3  *Ebook*

*For Mom and Dad*

# CONTENTS

*"Get all the education you can,
but then, by God, do something.
Don't just stand there, make it happen."*

—LEE IACOCCA

# PART 1

## EXPERIMENTAL MONKEY BUSINESS

# INTRODUCTION

———

An apple fell on the head of a brilliant physicist who sparked hundreds of years of research and experimentation that still scratches heads. Whoever said fruits weren't relevant to physics?

In America, instruction of theoretical physics in high schools and middle schools can be lacking. Ideas of modern physics like special relativity, dark matter, and just about anything you can mention that starts with "quantum" are typically studied by undergraduate and graduate students with an extensive mathematical background and determined interest in an empirical-based education. But academic tiers any lower than undergraduate or graduate in general don't learn theoretical physics in a classroom setting; for whatever reason, it either is not prioritized by curricula framers or too difficult for younger students to learn at the secondary level.

However, another reason for the lack of theoretical physics literacy amongst younger students might simply be that current approaches to relevant concepts are not accessible, enjoyable, and creative. A look at current literature might just scare students away, who might otherwise have appreciated the conceptual nature of the topics discussed. To students who have no explicit reason to hate physics, the conceptual nature of the subject can be enriching. The prevalence of ideas such as relative motion and particle physics, which can be shown to have many "relatable" applications in an everyday sense, may be more easily expressed without mathematics thrown in the mix.

Ever since I sat for my first physics lecture in electromagnetism, I found a certain mathematical complexity that I'd never discovered in previous classes. The fact that integration and complex numbers had meaning beyond the symbols used to denote them was a strike to the most sensitive part of my educational perspective, and it challenged me to study more of it. I discovered that high school students who might be interested in theoretical ideas could indeed study them. Rather than ask the question:

How can I learn theoretical physics?

They might be better off asking themselves:

How can I learn theoretical physics in a creative enough way for me to want to share my knowledge with others?

It certainly is true that mathematics aids in the exploration of physics amongst members of the research community, but such members dedicate themselves to a field that required a proficiency in mathematics. Not all high schoolers who might want to make this dedication in the future would be able to explore physics in time to make this dedication if they pursued the same studies that professionals do. It is a tricky loop: high school is an open academic environment in which students experiment with potential courses of study. If high schoolers cannot explore theoretical physics because the literature sources are heavily technical in that they impose ten years of mathematics coursework to be understood, then experimentation with theoretical physics studies would probably be difficult.

I reasoned that there had to be a way to enhance instruction of theoretical physics to appeal to a younger audience. Currently, fields like computer and data science are overtaking students' interests for financial, public networking, and biological applications. Where the discovery of the Higgs Boson and development of that discovery fits, I have no idea. Moreover, that's partially why I wanted to write Banana Physics: I wanted to have an idea. I wanted to encourage the pursuit of theoretical physics by young students like me before we are presented with the opportunity to drop it and instead pursue other fields. As much as I dislike the possibility emphasized in that latter point, I don't pretend to know the future.

When I started writing Banana Physics, my goal was to provide a "peek" at a collection of theoretical physics concepts. I intended this "peek" to be for students my age or younger and I wanted to write creatively. Certainly a sizable amount of physics literature provides ample resources for public reading, but how accessible that literature is to students my age, I have my doubts. In one of my favorite books, "Quantum Mechanics," author and Professor Leonard Susskind writes in his preface:

"The best way to understand physics is to write about it."

I knew that for younger audiences, this line had to apply differently. Professor Susskind writes about quantum mechanics (amongst other ideas) but he does so with mathematical prerequisites in mind. I read his book and decided that if I was going to write about physics, I wanted to make it more tangible for universal, "young kid" appeal. Admittedly, bananas were not the method I initially chose to develop said appeal. In fact, I can't quite say when exactly bananas entered the picture.

About one year ago, I took a look at all the physics content I had been exploring, keeping the mathematical rigor in the picture: Dr. Ramamurti Shankar's Open Courses, HyperPhysics, MIT OCW, Pauls Online Math Notes, and Princeton's online lecture notes. I worked through them piece-by-piece, formula-by-formula, idea that made no sense-by-idea that made no sense, and started to relate them to funny, quirky analogies. Interestingly, I found a way to describe many of them with some kind of banana analogy. First came electromagnetic wave theory, and then dark energy. The Standard Model of Particle Physics followed, then quantum gravity. I didn't think quantum mechanics would comply but it did, and it did so with bananas. Thus was born the messy manuscript of physics explained with bananas. Even though I tried to come up with a nice title for it, I was not as creative as I would have liked to be.

In this book, I take a look at eleven concepts in theoretical physics and place them in the context of quirky banana analogies. Some are a bit quirkier than others, some are shorter, some are longer, and some incorporate other things, like monkeys and rubber ducks. Each concept gets its own chapter (except for some, which are grouped together), but everything I talk about is accompanied with some kind of tangible, yellow fruit that provides a concise explanation of a very important question:

"What do physicists like Dr. Shankar and Dr. Susskind understand about the universe? And how can that understanding be communicated to aspiring physicists, like me, like my friends, and like other passionate kids who just want to learn some physics?"

In a nutshell, Banana Physics is my version of physics with creativity strings attached. I make the promise to you that you won't see any difficult calculus (or easy calculus, at that), linear algebra, or partial differential equations. There are no practice problems, no hardcore physics, nothing that should insinuate mathematical "scariness." Because it leaves these components out, this book cannot possibly teach you any physics. I think this is a fairly critical statement worth repeating: you will not learn physics in reading Banana Physics. You will learn reasons to explore physics by reading about it in unique frameworks. The book in your hands is not a "how-to," it is not a textbook, and it is most certainly not a creative version of any of the previously mentioned books. It is a collection of banana analogies that explain a collection of physics ideas. It perhaps could be your expectation, and certainly is my hope, that you obtain a creative glance at a marvelous playing field.

One of the biggest challenges of making progress in science is communicating it to others. Specifically, it is communicating science with people who aren't in the lab: professionals of other fields, teachers, and students.

Learning physics and science in general should not be forced. Nor does it require a narrow audience. How interesting would it be if a music, visual arts, or poetry enthusiast discovered something to explore further about physics after reading about it in an appealing and laughable language!

Banana Physics is a math-less, creative span of the space of theoretical physics. Intended for younger (high school) students, like you and me, to understand the basis of next generation physics in a more creative way. We start with an overview of light and matter, we proceed to Einstein, and we end with clashes between the two, other relevant areas, and experiments. It's an action-packed flying circus of banana silliness.

# 1

# THE BANANA SPLIT EXPERIMENT

———

## LIGHT

Depending on your experience and knowledge of physics, your answer to the question – "What is light?" could span many scales of creativity, ridiculousness, and logic. You might say it's "bright," it's "the Sun," or it's "whitish-yellow." But if I tell you to give me a physically rigorous description of light, your range of possible answers becomes narrower because you need to be more specific. It isn't enough to describe a colloquial characteristic of what you're trying to study and expect it to serve as a proper characterization. Perhaps you could think about what constitutes light, how light moves, what its properties are, and how it stores its energy.

When I say, "look at light," I don't mean look at the Sun. When you look at the Sun, you are seeing a glowing sphere of burning gases that consistently radiates light, but the sphere itself is not light. The light that is emitted from the Sun traverses the Universe in a certain trajectory to reach your eyes almost instantaneously, reflect back almost instantaneously, and render an image via retinal processes in your eyes and brain. For us to view images of matter, light must travel to and reflect off matter. This means that what we perceive light to do, illuminate objects in the atmosphere, is based on the idea that the dynamics of light can be modeled conceptually and mathematically. How light moves from place to place is important to an understanding of what light is.

If something moves, it necessarily follows that it has a specified composition. You can tell me that light is a continuous volume of "light fluid" and that this light fluid simply moves through space in the shapes of beams. You can also tell me, with equivalent hypothetical significance, that beams of light are composed of tiny composite "things." Meaning, light can be scrutinized on a fundamental level to be comprised of something smaller, just as you and I are comprised of organs, cells, atoms, and quarks. I'm sure you can come up with alternative explanations for what the light beam might be made of. I am not sure that I can address all of said imaginative underpinnings. I'll assume that all of your descriptions can be thought of as variations of the two ideas I provided above: light is either particulate or it is fluid.

A few examples illustrate why this generalization works. To perceive matter, one has to perceive the existence of a body that takes up space. The body could be made of anything that fills space, even if it is divisible. If the body were to be broken into millions of little bodies that fit together to fill the space, there would be no change to the amount of matter present. This is as if you placed fifty ripe, unpeeled bananas in one large, locked box. Because the box is closed from outside contact, there will be fifty bananas in the box no matter what you do to it. You can shake it, drop it, or hit it; the only change would be the form of the bananas. Perhaps they become so mashed that they resemble a banana smoothie. At that point, you would have made the matter into a fluid.

When physicists of the sixteenth and seventeenth centuries began to write down mathematical explanations of what light was, they employed similar reasoning. If our perception of light is predicated on its dynamics, its composition must also play a fundamental role. If it is composed of something, it might be a fluid or a collection of particles. Between these options, most people were relatively confused.

## THE DOUBLE SLIT EXPERIMENT

In 1704, Isaac Newton made a bold declaration and largely contributed to an expelling of the confusion. In his Principia, he described how light had a corpuscular nature. In other words, light was composed of very small entities, or particles.

The reason why Newton's contribution had (has) significance is not because he was the first physicist to decide between light being fluid or particulate. He actually contended against a pre-existing postulation that was made in 1687 by Christian Huygens. Huygens declared that light could not be made of particles and not only was it a fluid, it was a wave-like fluid. In the context of physics, a wave is an oscillating body of "something" that moves through a medium. Ocean waves provide nice examples because they physically match the definition but other waves, such as sound waves traveling through air, also qualify.

In general, there are two quantities that are very important to matter in physics: energy and momentum. For waves, both of these quantities are distributed across all points in the wave body. Particles are not like waves; unlike a wave, a particle's energy and momentum is concentrated in the region the particle occupies. When particles interact with other particles, there can be energy and momentum transfer, but the point of contact is the point of transfer. You take a banana and throw it at a wall, and the banana's energy and its momentum are both transferred at the point at which the banana touches the wall.[1] If you take a wave of water and throw it at the wall, the wave's energy and momentum are dumped at all points of contact on the wall, which means we have to work with a distribution of energy and momentum dumping.

Unfortunately, for Huygens, his wave theory of light was underdeveloped in the limelight of Newton's invention of Newtonian mechanics. Newton's laws of motion describe the motion of bodies in the universe and how forces between those bodies are related to said motion. And during roughly the same time in which he invented mechanics, he also invented the fields of differential and integral calculus to help him visualize what he was doing. The majority of physicists were heavily preoccupied in experimenting with Newton's laws of motion, which provided an entirely new framework for measuring energies, velocities, and forces in the known universe. People knew that if you threw a banana at a target, it would probably hit the target, but never before had the trajectory of the banana been quantitatively modeled to include such measures as acceleration of gravity, for example, which Newton precisely provided.[2]

---

1   Full transfers of momentum and energy are unlikely as a specific collision topology and environmental conditions must be observed for both of these quantities to be completely passed to the wall. As an approximation, you can assume complete transfer.

2   You can read up on Newton's work in just about any textbook on classical mechanics.

The reputation developed by Newton's other work was only one component for the popularity associated with the corpuscular theory but it was certainly compelling. However, in the late 1790s, a young English scientist by the name of Thomas Young hypothesized that Huygens' theory of light might be more strongly supported than could that of Newton. It was not easy at the time to experiment with light the way optical industries can do today, so Young's approach to this problem was simple. He took a light source (you can say a light gun) and placed it in front of a large metal plate. Behind the metal plate, he placed a detection screen, which is essentially a white screen that displays illuminated bands if light strikes it. If you had a similar apparatus and were to shoot light at some point on the screen, there would be a bright band at that point of contact. Where light does not strike are dark points.

If you were to hold the light gun, in front of you would be the metal plate and behind the metal plate would be the detecting screen. Young then inspected the metal plate and punctured a small slit in the plate. Although the slit was far too thin to allow a banana to pass, let's assume there are bananas that are allowed to slide through. If the light gun were configured to emit bananas and not light, when the gun is turned on, what is expected? For every banana that is shot out of the gun, there are two outcomes:

The banana hits a solid part of the metal plate and falls to the floor.

The banana passes through the slit and hits a section of the detecting screen.

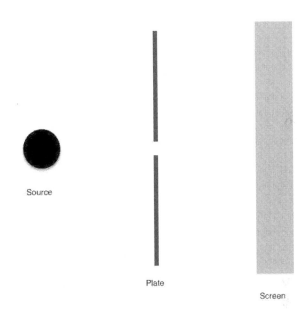

Source

Plate

Screen

You don't expect the likelihood of a banana making it through the slit to be equivalent to that of a banana hitting the floor. There is only one small slit but many more slit-sized sections of solid metal plate. It's like if someone covered a large wall with metal, blindfolded you, and asked you to run at the metal in hopes of making it through a small door carved into the surface. You know the slit path is unlikely but you don't suppose it is impossible, so there is a likelihood of you seeing a banana crash into the screen behind the plate. Where it hits the screen is of course dependent on the trajectory of the banana. It might go through the slit and be altered in such a way that it hits the screen a few centimeters off the screen's center. Or it might be perfectly unaltered altogether, which would result in the banana crashing into the center of the screen.

Dwell on this logic in the context of Young's experimentation with light. When many "beams" of light are sent at the slit, it cannot be expected for many of them to pass through the slit, but the ones that do will most certainly hit random areas of the screen. The marks would not be random enough to cover the full screen because realistically, most bananas would simply be displaced by a small amount and end up hitting the screen off-center. The final result should simply be a screen with a random collection of center-concentrated bands. If Young saw this, he would have had to consult a different experimental protocol to reformulate Newton's theory, because this is supportive of the corpuscular theory of light. If light is composed of tiny, packaged particles that move in a unitary fashion, it must take the trajectory of any other particle that could be specified to represent, for example, a banana.

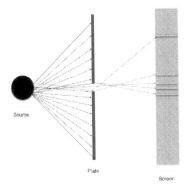

What you would expect to see, if light traveled in straight beams. The majority of the light beams (underemphasized in the diagram) would simply crash into the metal, but a select few would find their way through to the screen. You might expect some of these light beams to hit off center, but most of them will illuminate the center of the screen.

Source

Plate

Screen

When Young shot the light at the metal plate, he saw something very peculiar. On the screen was a band distribution. Every section of the screen was illuminated to some extent and not a single section was left unlit. The center of the screen was the brightest but fanning from the center were bands that became less bright as the distance from the center increased. If one were to draw a mark on the right side of the screen, such that the distance from the screen signified how bright the corresponding section was, as shown below, one would obtain a curve. Of course, one would have limited knowledge of how that curve came to be.

If Newton's theory is held as reliable in interpreting these results, the only explanation to be given is that the theory holds and Young's experiment is flawed. For example, as rays of light passed through the slit, they could have been deflected by the edges of the metal plate. This deflection might have caused the light to travel to all sorts of locations on the screen, which explains the observations.

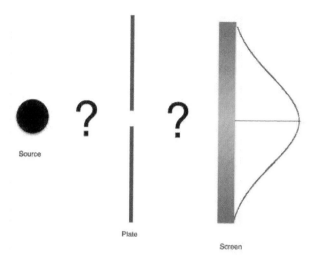

Unfortunately, this is not a pragmatic explanation because while no experiment is perfect, it is highly unlikely that an experiment can be so perfectly imperfect. Light could be deflected inside the slit to hit different areas of the screen, but can it be said with a straight face that there was enough deflection, in a precise enough fashion, to produce an even distribution on the screen? That this supposed deflection of light caused it to illuminate every part in symmetric fashion? It would violate our sanity to say something like this.

Young was able to look at this behavior in the context of something called diffraction. Diffraction happens when a wave approaches a slit as a moving wall, or plane wave, that is parallel to the orientation of the slit. As this plane wave moves into the slit, it is diffracted into semi-circular wave fronts. What was a plane wave became a collection of semi-circular waves of different radii (size), but share the same center. Young therefore proposed that light emitted from the source, approached the plate as a plane wave, and broke into semicircles as a result of diffraction. As these semicircles start to approach another surface, such as a detecting screen, there would be interesting observations because semicircles are not particles. There is one point on the semicircle that is the highest up and infinitely many points that are below it, with two being the absolute lowest at either end. Imagine a semicircle of light approaching the screen. There is one point that will reach the screen first, which is the top point of the semicircle, but every other point will reach later on. Assuming the center of the screen is aligned with the top point of these semicircles, the first point of semicircle-screen contact will be the point at which the most light is dumped. As you look outwards from the center of the screen, there is less brightness because the "height" of the semicircle becomes smaller. The "height" of the semicircle is referred to as the amplitude of light.

If you have prior knowledge of wave physics, you know that amplitude is a characteristic of waves. When a wave oscillates in space, it moves from some base point to its highest point, from the highest point back down to the base point, from the base point even further down to the lowest point, and from the lowest point back up to the base point. Usually, a wave also moves through the space itself, which means there is not simply one base point, low point, and high point; these points each lie along one line that describes the motion of the wave, both as it oscillates and as it moves. The distance between base to high is the same as it is between base to low, and this distance is referred to as the amplitude. The frequency of the wave is the number of cycles it can make (that is, from base to high, high to base, base to low, and low to base) in one second.

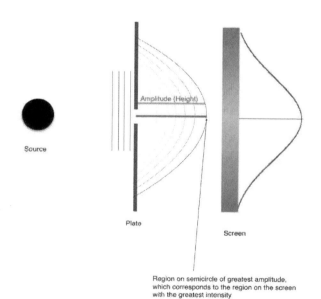

Source

Plate

Amplitude (Height)

Screen

Region on semicircle of greatest amplitude, which corresponds to the region on the screen with the greatest intensity

You can think of the semicircles as wavelike fronts. Young reasoned that the brightness, or intensity, of a band on the screen was proportional to the height, or amplitude, of the point on the semicircle that was approaching it. Greater amplitude would mean a greater intensity on the screen, which fits with the model: the point of greatest intensity was the center of the screen, which happened to align exactly with the point on the semicircle of greatest amplitude, or the point highest up on the semicircle. As the amplitude decreased, so too did intensity, which explains why the distribution looks as it does.

If light breaks into semicircles, semicircles have amplitudes, and amplitude is a property of waves, then at first inspection light should probably be a wave for the assumption of diffraction to hold. Young reasoned that there was still further evidence that could potentially support his claim of light being a wave. He made a second slit.

Why should two slits give anything unexpectedly interesting? You can probably give me a very intuitive explanation as to what Young would see had he opened two slits rather than one: if one slit is open and gives an intensity distribution on the screen, opening another slit would simply amplify the intensity distribution.

Let us say you took a banana and placed it on a table in a dark room. You then take in your hand a flashlight that you shine on the banana. If I tell your friend to take out another flashlight and to move one meter to the right of where you are standing, what do you expect to happen if both of you shine your flashlights on the banana at the same time? You expect the banana to be brighter than it was when only you were shining your flashlight on it. It certainly fails intuition if the banana does not get brighter, or worse, if the banana becomes less bright.

It would also fail our intuition if any region on the screen became less bright because the screen in the experiment is analogous to the banana. You cannot introduce a second source of brightness to illuminate the screen and expect the screen to become darker. Any given region would also not be dark, because you cannot "cancel" light by bringing more light in. The only common sense observation is an additive intensity distribution of both slits, meaning no matter where on the screen you look, two slits will make it brighter than one.

Except this is not what the screen looked like at all. With two slits open, the center of the screen was still the brightest, and fanning from the center at regular intervals were bright parts, but in between those bright parts were dark parts!

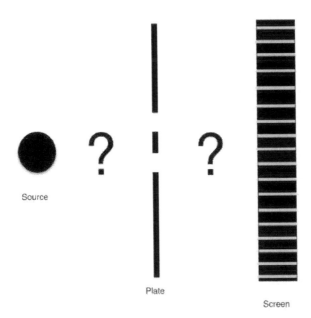

Source

Plate

Screen

## WAVE PHYSICS: A FEW WORDS

Let us say you have a wave emitter that can send plane waves at the plate. The emitter emits one plane wave that approaches the slit. Upon passing through the slit, the plane wave diffracts into two sets of semicircles called set A and set B. Pretend that B is temporarily absent. If A goes towards the screen, you expect to see a similar intensity distribution compared to the one produced with Young's light passing through one slit. If you like set B better and you want to get rid of set A, you still see the same intensity distribution but it is merely shifted to a different location on the screen.

When two waves collide, they don't simply crash and move in opposite directions. If you throw two bananas at each other, you will see a wide variety of results depending on the spatial orientation of the collision. They might move in opposite directions, they might simply fall to the ground, one could go flying, and the other could fall, or vice versa. There are many possibilities, as I am sure you are aware.

Except waves are not bananas. When you throw waves at each other, they interact and form hybrid waves. These hybrid waves are called resultants and they have interesting properties depending on how the collision between the initial two waves happened. If two identical waves have an amplitude of one meter and collide, the resultant wave will have an amplitude on a range of values, the minimum being 0 meters and the maximum being 2 meters. Why is this so?

The amplitude of the resultant may be amplified by two waves when they collide but the collision must happen at a specific point known as the antinode. The antinode is where full "collaboration" between colliding waves happens; the resultant amplitude is the sum of the amplitudes of each wave. The node of the collision, on the other hand, is the location of minimum collaboration, meaning the resultant amplitude will get as close to zero as possible. When waves collide at the antinode, they undergo constructive interference, such that if each has amplitude of one meter, the resultant will have amplitude 1+1 = 2. When waves collide at the node, they undergo destructive interference, such that if each has amplitude of one meter, the resultant will have amplitude 1−1 = 0.

Often, most colliding waves do not have identical amplitude, which means complete constructive interference may be observed, in which for example one wave of two meter amplitude collides with another of five meter amplitude to produce a resultant of seven meter amplitude, but because the waves have different amplitudes, full destructive interference cannot be observed. However, in the context of the double slit experiment, both slits are taken to be the same size, which means the sets of semicircles are also the same size and have the same range of amplitudes. When these sets collide with each other (they have to because they pass through a small space in the experiment), there are most definitely points of constructive interference in which one would see more amplitude than what one would have seen with one set. But there are also points of destructive interference, in which one would see zero amplitude. Zero amplitude means zero intensity on the screen, which means a dark spot.

Constructive and destructive interference are variations of a phenomenon known as interference, and Young proposed that the semicircles of light that intersect with each other in this experiment also interfere with each other. The points of destructive interference are the points of zero intensity, which explains the counterintuitive observations made prior.

Light therefore diffracts into wave fronts through the slits, at which point it is well-positioned to observe interference effects in the region between metal plate and detecting screen. In accepting this statement, one is forced to acknowledge the inclusion of two well-reasoned properties of light's behavior in this experiment that match exactly with two properties of wave behavior: diffraction and interference.

From 1801 to 1803, Young took his observations to the Royal Society in England and presented a series of lectures that proposed the first experimentally supported theory of light: the wave theory.

## ENTER ELECTROMAGNETISM

The fundamentals of electricity were proposed by Michael Faraday, who in 1839 developed the first interpretation of the electrostatic and eventually electromagnetic force. Far before Faraday, the phenomenon of electricity had already been discovered, but what Faraday did was unique: he suggested a potential unification of electricity with another phenomenon known as magnetism.

Magnetism works differently than electricity, but there are still magnetic fields that permit the exertion of magnetic forces on charged, moving particles in the field. Simply stated, he found that if you take a metal wire and produce a magnetic field inside of it, you can vary the magnitude of the magnetic field and induce a current in the wire out of nowhere. Typically, you can connect a wire to a power supply, which drives electrons around in a circle and thereby produces a current. This is in fact the definition of current: the amount of charge that moves around per unit time. In Faraday's time, everyone knew how to produce a current in a wire but nobody expected it to be related to magnetism. It was a completely different force that didn't seem to bear any relation to electricity, but according to Faraday, it most certainly did. Changes in a magnetic field produces a current, which means magnetism induces electricity.

The follow-up question for Faraday was obvious: if magnetism can produce electricity, can electricity do the same for magnetism, and if so, how? This went unanswered until 1873, when James Clerk Maxwell published his equations of electromagnetism. Maxwell is widely regarded as the father of electromagnetism, as his equations explain how magnetic fields induce electrical currents and how those induced currents induce more magnetic fields. Maxwell stated that the electric field and the magnetic field are both orthogonal (perpendicular), oscillatory fields that move together through space. Maxwell proposed that together, these translating fields constitute what is referred to as the electromagnetic wave.

An electric field traveling through space as a wave.

A magnetic field traveling through space as a wave. If you were to stand above this wave, it would appear identical to the view of the electric field as shown above.

Electric and magnetic fields are put together in an orthogonal, or perpendicular, orientation. This combination forms an electromagnetic wave.

Given properties of an oscillating system such as the electromagnetic wave, one could calculate its wave velocity. Maxwell calculated it to be a precise $2.998 \times 10^8$ meters per second.

Experiments since 1676 followed the initial observations of Ole Römer when he calculated the speed of light to be roughly 300 billion meters per second. The speed of light was verified for nearly one hundred years. When Maxwell calculated the same speed for that of an electromagnetic wave, the connection to be made was incredibly profound.

Young's wave theory was not only supported but also deserving of resonance beyond what was initially thought reasonable. It became apparent that light might be an electromagnetic wave, that it propagated when electric fields induce magnetic fields and those magnetic fields induce more electric fields, and that it moved at a very unique, fast speed of $2.998 \times 10^8$ meters per second. In no experiment has an object been identified that can travel at the speed of the electromagnetic wave or of light, meaning the profundity of their relationship is predicated on the exclusivity of what physicists have studied about the universe. $2.998 \times 10^8$ meters per second even has its own variable: c. Nothing else moves faster.

There is a lingering question to be addressed: if physics is supposed to account for "how stuff works," and if Thomas

Young is right about interference and diffraction in terms of light's behavior, then why doesn't the banana and flashlight example work? Why is it that if you use a banana and some flashlights, you do get an additive intensity distribution on the banana, but if you use light in the double slit experiment, you still get what Young got?

There are experimental nuances that make the interference and diffraction effects possible. The first is that in order for light to interfere as it did in the double slit experiment, the slit through which it passes must be very small. This is because individual segments of the electromagnetic wave are microscopically small and if you try to shoot light into a double slit experiment, you must make sure you isolate some finite section of the wave. If you blast the slit with high intensity light, the plate will probably explode.

Another nuance is the fact that in the double slit experiment, the distance between the two slits must also be microscopically small because that distance must be essentially the same as the wavelength of the light you use. If the distance increases to orders of magnitude larger than the light's wavelength, the experiment fails. And because the wavelength of light generally is nanometers in length, the distance between the slits must also be nanometers in length. In the banana and flashlight example, you likely won't stand nanometers away from your friend. There is something called personal space, you know.

Moreover, the final nuance, perhaps the most important nuance, is that the light that you shoot at the metal plate has to be of a defined wavelength; otherwise, you can't guarantee that the "semicircles" it splits into can destructively interfere to give an amplitude of zero. In order to measure the interference effects as perfectly as they were in Young's experiment, you need to make sure the light you are using is homogeneous. Unfortunately, this is not something you can guarantee with your banana and flashlight example. You can invent a flashlight that gives light of one amplitude but this doesn't mean there is no other light in the room you are in. There is light that shoots around everywhere in our atmosphere from various sources, and all can be and are different in a number of ways; frequency, amplitude, wavelength, etc. We are being hit with light at all times even if we don't feel it. But experiments are not as tough as we are. Experiments are precision-based and require a certain set of conditions to give accurate, verifiable results. You cannot guarantee experimental conditions in a non-experimental environment, even if you are careful. Using an apple won't help either.

# 2

# CATCHABLE RUBBER DUCKS

---

It suffices to say that the double slit experiment introduced a revolutionary understanding of the dynamics of light. The observations of diffraction and interference hinted at potential flaws in previous Newtonian literature, but the change did not settle as soon as it appeared. Defendants of the corpuscular theory of light, which described light as a collection of particles, continued to defend, despite the fact that challengers found new ways to challenge. Ultimately, it took the foundation of a seemingly unrelated field of physics – electromagnetism – that provided enough of an intuitive framework for the existing empirical complement.

The foundations of electromagnetism can and do occupy several books, so I will not attempt to explain Maxwell's work in this chapter. What you should understand is that electromagnetism is an idea that, when first written down, brought together previous discussions of electricity and magnetism into one force. Before Maxwell, electricity and magnetism were understood to be distinct, unrelated natural phenomena. The famous "induction coil" experiment showed that by simply applying a magnetic field to a coil could scientists observe a current in the coil that was previously nonexistent. The relationships between electric and magnetic fields are elegantly summarized in Maxwell's equations.

The importance of Maxwell's work to discussions concerning light being a wave revolves around velocity. Maxwell understood that electromagnetic waves form when an electric field is sent oscillating, perpendicular to an oscillating magnetic field, that together move through space. Through theoretical calculations, Maxwell found that the speed of such an electromagnetic wave is approximately 300 billion meters per second. If you have studied physics, you might have already noticed that this is also the speed of light. Experiments from as far back as 1676, when Ole Romer first calculated the speed of light, point to this simple correlation. Moreover, it is the very simplicity of the idea – that speeds match up – that contrasts the momentous effect Maxwell's, Romer's, and Young's work had on humanity's quest to understand what we don't, and just might, understand tomorrow. Once these details came together, it was very difficult for any twentieth century physicist to nullify the wave theory of light and attribute it to random experimental behavior, miscalculations, or just physicists being crazy. Light was a wave.

Nevertheless, despite the glory, challenges found their way through. The problem is that physics doesn't stop at dynamics, because while the study of motion of objects is important, it doesn't span our understanding of them. For example, you can send a light wave crashing through space and talk all you wish about its amplitude and its frequency, but what about its energy? What about its momentum? Physical observables such as energy and momentum are extremely important to the systems in which they can be studied. Certainly, light carries energy, but with wave theory, the details aren't clearly defined.

You are familiar with amplitude and frequency, but here is a recapitulation if you forgot: a wave moves through a medium, but it also oscillates in the medium. It has some base point from which it starts, it travels from base to high, it goes back down to base, goes from base to low, from low back up to base, and it repeats. The amplitude is the distance from base to high and from base to low (they are the same distance). Frequency, on the contrary, is not a measure of distance. Frequency measures how many full cycles the wave completes in one second. A guess about light's energy would probably sound something like this: "I know the electromagnetic wave has two important properties, frequency and amplitude, so energy might be related to either of these or both of these." Yet, we need more than just a guess. And after the double slit experiment, this is exactly what physicists strived to obtain: a more specific understanding.

## THE PHOTOELECTRIC EFFECT

What was postulated seemed simple. The energy of an electromagnetic wave, or the capacity of the wave to "do something," should be proportional to how strong the wave is. This was not a quantitative description, so the next step was to define a metric for "strong." Intuitively, the measure of the strength of a wave relies on how much it hurts you when it hits you. When you go to the beach and stand at the shore, you understand that the bigger the wave is, the higher the likelihood you're getting knocked down. If you see a twenty-foot wave, it does not matter how many times the wave hits you, because it only needs to hit you once for you to fall. What this means is that how many times the wave hits your face is not a relevant consideration if the height, or strength, of the wave is sufficiently strong enough to knock you down. The frequency of the wave – or how many times it hits you in a given time interval – is not related to the wave's energy, because you could have fifty waves hitting you in one minute, but if they are all four inches high, you will triumph. If those waves were each twenty feet tall, that's a very unfortunate situation. Please don't try this.

In 1887, a physicist by the name of Heinrich Hertz performed an experiment that tested what was called the photoelectric effect. Hertz was someone who wanted clarifications about energy. If he could somehow prove that light's energy is solely a function of its amplitude and is subsequently irrespective of its frequency, the wave theory would again collect a feather in its hat.

Let's say you are given a banana that is hard to peel, and you only have one minute to peel it. You have two options: you can ask 500 two-year-olds to try peeling the banana open one by one, or you can ask one bodybuilder to do it. You recognize that the number of people who work per minute is much greater if you opt for the five-year-olds, but if you choose that option, you understand that each of these two-year-olds has the same amount of muscular capacity in the action of peeling a banana, and it is more than likely that it's not enough. If the banana is hard to peel, why would you expect multiple five year olds, with independent attempts, to complete the task? Opt for the bodybuilder; it doesn't matter if only one person per minute is working, because the strength to peel the banana is likely there.

Similarly, if light were to "do" something, like peel a banana, how many tools it uses per minute would be irrelevant, as long as you have strong enough tools that can do the task sufficiently. This is Maxwell's reasoning: the energy contained in the electromagnetic wave is dictated by its strength. In the context of wave theory, how "strong" the electromagnetic wave is corresponds to amplitude. How many people working corresponds to frequency.

You might ask: "why can't you ask five hundred body builders to do it?" That's good reasoning, but realistically, think about what would likely happen. We know that light's energy is some sort of function of its strength, or amplitude. However, we do not know if light's energy is in any way related to its frequency, meaning a high frequency, large amplitude electromagnetic wave could have the same energy as a low frequency, large amplitude electromagnetic wave. Five hundred body builders, in the context of energy, have the same strength of one body builder. This is counterintuitive, but if the task is simply to peel a banana, and each body builder in the group of five hundred will give it a go separately, then the first body builder's attempt is sufficient. You don't need the other 499 body builders. However, Heinrich Hertz didn't rely on intuition. His experiment was as follows:

Hertz took two metal plates and put them side by side. Metal is composed of metallic atoms, but shared between the atoms are what's called delocalized electrons. These electrons essentially sit inside the metal and don't do anything, unless they are given some sort of energy to get up and start moving. If they are given a small amount of energy, they only reach an excited electron state and stay inside the metal, but as more energy is supplied to the electrons, they can travel to the metal surface and "jump" off.

Hertz's reasoning was that if he brought the two plates close to each other and shined light on only one of these metal plates, let's say plate P, then he could force electrons to go from plate P to the other plate. We'll call this other plate N. If light crashes into plate P and gives it enough energy to send its electrons to plate N, then the two plates eventually change in electrical charge. What happens? Metal is typically electrically neutral, meaning it doesn't carry a net positive or negative charge. However, if electrons leave plate P, then negative charge is ripped off plate P, which means it will be positively charged. These electrons jump to plate N, on which negative charge is being dumped constantly. Plate N becomes negatively charged, and plate P becomes positively charged.

When light shines on plate P, it doesn't simply give all of its energy to electrons for them to start moving. There is a specific requirement that the light must have met before ripping any electrons off, and this is called the work function. All electrons have a work function that dictates how much ɟy it takes to rip it off the atom it is orbiting. However, if the light that shines on plate P had more energy than what is defined by the wave function, the electrons have extra energy. This "extra energy" is energy of motion, or kinetic energy, and it is why electrons are sent scurrying from plate P to plate N if the energy of the light is sufficiently large.

If Hertz could somehow measure how much kinetic energy these electrons have, he might be able to define how much energy the light had, except, there are no "energyometers" that can be directly applied to metal atoms to measure how much kinetic energy migrating electrons have.

Instead, Hertz took these two plates and connected them in a circuit with two other components: an ammeter and a power supply. Hertz realized that if electrons continued to accumulate on plate N, he could force them to move outward into the circuit. This is why the power supply is there: the power supply provides a voltage, or electromotive force, that encourages electrons to move through the circuit. As electrons move into the circuit from plate N, and if the circuit is designed correctly, the electrons can be made to cycle around back to plate P. And because the power supply is still pumping voltage in the circuit, electrons continue to move from plate P to plate N, and the cycle continues.

The reason why this experiment makes sense is because of the ammeter. The ammeter's job is to measure a current, which is essentially defined as the amount of electrical charge that passes through an area in a given time frame. Electrons are electrical charges, so if electrons are traveling in a circuit that contains an ammeter, the ammeter will detect a measurable current in the circuit wires. If Hertz could send light onto plate P, watch electrons migrate to plate N, cycle through the circuit, and give current readings in the circuit, then he had a decent picture: by reading current values from the ammeter, he could understand how many electrons were traveling in the circuit, and with which energies. In addition, because the electron energies came from light, the ammeter current readings were closely related to the energy of the light that shined on plate P.

Hertz could not simply obtain a light source that had energy settings built in, because if such technologies did exist, then this experiment would have been pointless. However, he could use a light source with frequency and amplitude settings, and that's exactly what he did. By establishing some continuum of amplitude values to test, and establishing a parallel continuum of frequency values to test, he could observe if a high amplitude really did mean a high energy (if frequency really did have no significance).

He set the amplitude to a very small value with the frequency kept small as well. No current, exactly as predicted! Small amplitude, as stated by Maxwell, yields insufficient energy, so much so that electrons cannot make it out of plate P. Because no electrons leave plate P, no electrons travel to the circuit, which means no current. Next, Hertz raised the amplitude a little bit, but kept the frequency the same. Again, no current. He continued to raise the amplitude until he could not raise it any further (again, frequency is constant and small) and there was no current.

Most proponents of the wave theory expected light's energy to be positively related to its amplitude, which means if we increase the amplitude, we increase the energy. However, the amplitude of light, as dictated by this experiment, did not have any significance in terms of its energy. As Hertz continued to increase the amplitude while maintaining the frequency at a tiny value, there was no ammeter reading. Electrons were not liberated from plate P.

Then as the frequency of the light source slowly increased, Hertz found that the current in the ammeter also increased. If the frequency was set to its maximum, the current was also at its maximum, and if the frequency was zero, so too was the current. Even if the amplitude was kept tiny, small, intermediate, or massive – as long as the frequency was sufficiently large, there were electrons moving in this circuit.

## REVISITING THE CORPUSCULAR THEORY

So what does this all mean?

Immediately, nobody understood for certain. It took the brilliance of Albert Einstein, who put it all together with his 1905 paper, Concerning an Heuristic Point of View Toward the Emission and Transformation of Light. Einstein proposed that light could not be a wave, at least not in the context of its energy profile. If the frequency of light dictated its energy, how much light on the metal would be critically more important than how strong the light was. That means light is somehow quantized; it was composed of entities that each provided a finite energy. More of these entities of a given size "S" hold more energy than a single entity with size "much bigger than S." No matter how large the amplitude of light was, if the frequency was not sufficiently strong, it did not have enough energy to rip the electrons off Hertz's plates. From there, Einstein wrote down the famous relationship: The energy of each unit of light, or photon, was the product of its frequency and Planck's constant.[3]

If you are relying on intuition, contradictions should be floating in your mind. Imagine what would happen if you

---

3   Max Planck was a German scientist who experimented with a phenomenon known as black body radiation a little before Einstein's time. The reason why the constant is named after him is because he was the first one who proposed that light's energy was a linear function of its frequency.

stood in front of a large ocean wave and let it hit you. What would you feel? Surely, the wave does not choose one spot of you to hit; it spreads across your entire body and deposits its energy and momentum on your torso, arms, legs, and face. However, if someone else threw a rubber duck at your face, the story is different. The rubber duck will hit one part of your face, and its energy and momentum is transferred at that part. Where it hits is irrelevant, because for all possible trajectories it could take, the end result is a localized collision with one point. In other words, rubber ducks, like particles, are localized in space. They are not like waves, which are spread and oscillate in the area of spread. You cannot catch a wave, but you can catch a rubber duck.

Despite the findings of Thomas Young and James Clerk Maxwell, who seemed to conclusively show that light is a wave, Einstein, Hertz, and even Newton collaboratively suggested that existing work was not so conclusive after all. The core argument was energy: according to wave dynamics, the energy of a wave should be proportional to its amplitude, or how strong it is. Einstein's explanation of the photoelectric effect was a stark (and empirically justified) contradiction. Frequency was instead the central metric, which meant that light could not possibly be a wave if energy arguments were to be respected. Physicists were confronted with a momentous problem: here was team Young and Maxwell who said you can't catch light because if you could, it wouldn't display interference and diffraction, and it wouldn't move at the speed of an electromagnetic wave. On the other side of the table was team Einstein and Hertz, who showed that you can indeed catch light just as you can catch a rubber duck, because its energy wouldn't be proportional to frequency if light, or rubber ducks for that matter, weren't catchable.

At first glance, the issue of wave-particle duality might not seem like an issue. Why should it not work? If people knew that describing light as a wave agreed with the mathematical behavior of its momentum, and that describing light as a particle agreed with the mathematical behavior of its energy, then why should anybody have any reason to be concerned? Energy and momentum are after all distinct physical phenomena. But as we will see, to the majority of physicists since Newton's time, this is neither an acceptable nor a comfortable approach. If the physics of light accounts for its behavior in experiment in a certain sense, it should account for its behavior in all experimental situations. That means light can't be described as a rubber duck that can be caught and an ocean wave that you can never catch. This is the quest for systematic universality of physical interpretation, and it does not end at wave-particle duality.

# 3

# TINY BANANAS:
# SIMPLE PLAYS OF A
# COMPLEX GAME

---

In the middle decades of the twentieth century, a very charismatic professor by the name of Richard Feynman charmed the world of physics with breakthrough discoveries in electromagnetism, nanotechnology, quantum information science, and particle physics. Along with a number of publications, award-winning research projects, and a Nobel Prize, Feynman is known for saying,

"Nobody understands quantum mechanics."

I believe the best advice I can give you before you read this chapter is to keep an open mind. Your intuition will act as a fence to the rebellious teenager who wants to throw toilet paper on your lawn, which is what quantum mechanics is. You can't do that. The teenager will make it over the fence whether you like it or not.

You can also imagine quantum mechanics as a banana, with a beautifully colored skin that represents conceptual complexity. Behind that skin is a dense network of mathematical conjectures that redefine the meaning of intuitive observation. It doesn't matter where you look; the most you can do is to wish I had just said orange.

What I will attempt to do in this chapter is tell you about experiments that have been conducted and explain the models that have been formulated to fit the results. If you end up loving what you read, I strongly encourage you to visit websites and books to pursue deeper learning (that means with math). If you end up not liking it so much, well, that's harder to address. But in any case, do not feel disheartened if it makes no sense to you. You aren't alone.

## A CURIOUS QUESTION

It was known since roughly the 19th century that light observes wave-like properties in certain experiments, meaning in the context of these experiments, we can affirmatively conclude that light is a wave. However, if we go off and do different experiments on light, we can employ entirely different reasoning in stating that light is in fact a particulate collection of tiny photons. The good news is that we have two reasons to believe light is "something," the bad news is that they're intuitively conflicting answers. We have no other choice but to make the crazy conclusion: light demonstrates wave-like and particulate behavior. If you look at the history, something scientists first thought was a wave was also a particle, but here is the question history still has: does that necessarily mean that what we once thought was a particle, is also a wave?

In 1924, Louis de Broglie wrote down two equations that answered this question. He postulated that particles may be described by what is referred to as matter waves. A matter wave is a mathematical idea that essentially describes particles as waves. Intuitively, we understand that particles and waves can't be in simultaneous existence, but de Broglie said that matter can be both at the same time.

We should first understand what a "particle" is in the context of quantum mechanics. The classical particle everyone is familiar with is the atom. Atoms are known to constitute all elements, which constitute all substances that we interact with: pillows, wood, stainless steels, and rubber ducks are all made of some collection of very small atoms. We have no problem with defining atoms as particles, but the problem is, quantum mechanics calls for very specific experimentation. And in specific experimentation, you need to employ statistical methods in designing your experiment in order to obtain verifiable results. One of the most basic concepts in experimental design is finding a representative sample of whichever population you are working with. For example, if you want to assess annual academic performance of fifth graders in Texas, you don't test twenty-year-olds living in Maine, because that isn't a representative sample. Even if you give the Maine twenty-year-olds the same test you would (should) have given the Texas kids, your experiment unfortunately does not address your initial line of questioning. You had better work with Texas fifth graders to understand how Texas fifth graders behave.

Unfortunately, it's very difficult to find a representative sample for all atoms in our universe, because atoms are very inconsistent. Different elements are composed of specific atoms that have a fixed number of nucleons (protons and neutrons) and electrons. In table salt, you have two elements known as sodium and chlorine, and those elements are composed of sodium and chlorine atoms. It isn't just the number of neutrons and electrons that differ; they react differently to acids, they behave differently in solutions, and they are used for very different industrial purposes. If you want to ask a question of the form: "how do particles behave?" you can't use atoms, because bringing atoms into experiments subsequently introduces a number of variables that are too difficult to deal with. So we need to do better than atoms. We need something that exists in the same fashion across the entire universe.

Electrons are subatomic particles that exist in atoms. You might be tempted to apply the textbook definition that describes electrons as orbiting the nuclei of their parent atoms. Spare yourself the hassle. Think of an electron as a point particle in space. We can take electrons to be the particles in quantum mechanics experiments due to their consistency. If some guy on the street tells you that for ten bucks, he'll give you special electrons that can tap-dance, you know you can't be fooled, because as long as you have your own electrons at home, you know electrons can't tap-dance. Contrary to atoms, electrons are experimentally convenient and practical.

So now, we arrive at de Broglie's hypothesis, only a bit more specified: electrons observe wave-like properties, and can thus be considered matter waves. This hypothesis was formulated in tandem with two equations, known as the de Broglie relations:

$$E = hf$$

$$p = k(1/\lambda)$$

Where "E" is the energy of the electron as a particle; "f" is the frequency of the electron as a wave; "p" is the momentum of the electron as a particle; "$\lambda$" is the wavelength of the electron as a wave; "h" and "k" are constants. You can use these equations to relate relevant properties of electrons that apply to both wave and particle definitions. In doing so, it becomes apparent that for matter to exist simultaneously as a particle and a wave is not only shown mathematically in these equations, but a requirement for these equations to hold. This was the premise of the matter wave idea. Particles, which we don't suspect to be continuous disturbances through a medium, are also waves. Their properties as waves can be written in relation to their properties as particles, and vice versa. As illogical as you might think they are, these equations are by no means a mistake on de Broglie's part.

The majority of physicists at the time did not agree with de Broglie, because electrons were known to be particles, and an extension of wave-particle duality of light could not possibly apply.

## CRYSTALS

In 1927, two physicists at Bell Laboratories, Clinton Davisson and Lester Germer, conducted an experiment that supported de Broglie's hypothesis. Davisson and Germer took an electron gun and shot some electrons at nickel crystals. Nickel is what's called a transition metal, and it has a certain crystalline density that allows for a phenomenon known as electron scattering. According to these physicists, upon shooting electrons at a nickel crystal, you would expect those electrons to simply crash and bounce off. That's what happens when particles collide with surfaces: they bounce off. If you put a screen around the nickel crystal and forced the reflected electrons to crash onto the screen, you can use a fluorescent marking method to track where exactly they crash. In essence, you sort of have a deflection screen that models exactly where electrons go after they bounce off the crystal, not identical, but similar to the one you had in the double slit experiment.

Except, that doesn't mean you'd see double slit results. If we hypothesize that electrons are particles, we should not expect them to do anything else but bounce off and hit the screen randomly. Unless you lose your sanity, it's very difficult to imagine an electron that observes diffraction and interference after bouncing off a crystal. The good news is that we can keep our sanity for now – Lester and Germer didn't see diffraction or interference. The other good news, at least for de Broglie, is that they did see something else.

An X-Ray is a wave used in biomedical applications. X-Rays are not particles; the idea of wave-particle duality extends to electrons and to photons, meaning a detector that emits X-Rays is a detector that emits waves. There is no funny business concerning this assertion. Imagine a detector that emits X-Rays. When these X-Rays are shot at crystalline solids, there are four things you can measure. Two of these things are properties of the crystal itself and can thus be considered irrelevant as long as we can keep them constant. The other two, however, are important. They are the wavelength of the incoming X-Ray and the angle of scatter after collision. As an X-Ray begins to crash onto a crystal, it has some measurable wavelength, because it's a wave! When it reflects, it does so with an angle of scatter you can measure from some normalized baseline. These quantities are related as follows:

$$n\lambda = 2d\sin(\theta)$$

Where "$\lambda$" is the wavelength of the incoming X-Ray, "d" and "n" are the variables of the crystal you don't need to worry about, and "$\theta$" is the angle of scatter after the wave crashes into the crystal. As a reminder, what is meant by the wavelength of the wave is as follows: if you imagine a wave traveling along a straight line, where the straight line is the base point for the oscillations, then the wave starts at that line, goes up, comes back down to the line to go the same distance further down, and then comes back up to the line.

The difference, however, is that the final point after the full oscillation is farther away from where it started, meaning there is a distance between the initial point and the final point upon completion of a full oscillation. That distance is referred to as the wavelength. Waves can have different wavelengths, and Lawrence Bragg in 1912 took a wave emitter that allowed him to vary the wavelengths of X-Rays. He then sent them crashing into a crystal and measured the angle at which they scattered off. He found that the wavelength of the incoming wave was a sinusoidal function of the reflected wave's angle of scatter. That sinusoidal relationship, the one described by the equation above, is known as Bragg's law. The relationship between Bragg's law and quantum mechanics is critical.

Davisson and Germer reasoned that if X-Rays were waves that obeyed Bragg's law, then waves in general might also obey Bragg's law, and because Bragg's law is a mathematical relationship and not a postulation, you can predict, with relatively high accuracy, the angle of scatter of any given wave after you shoot it at the crystal.

However, if you shoot an electron, and maintain that an electron is a particle, those scatter angles should be random. If they observed some sort of pattern, it might reveal a relationship between the electron and the crystal that might call for further explanation. If that relationship were to be in accordance with Bragg's law, you don't just have a result; you have a revolutionary discovery that changes a field. And that discovery is the conclusion de Broglie already made, the one that asserts the (seemingly) intuitively impossible, that electrons are indeed waves.

In order for a wave to obey Bragg's law, the wavelength of the wave as it moves toward a crystal can be put into Bragg's equation. If you do the math correctly, you get an angle of scatter from the equation that can be compared with the angle of scatter you measure in the experiment. If both angles match, the wave obeys Bragg's law. As long as you have the wave's wavelength, you can assert if Bragg's law applies. If your goal is to shoot an electron at this crystal and assess how good a job Bragg's law does to predict the angle of scatter, you evidently need a wavelength for the electron, and interestingly, that implies the use of de Broglie's relations, stated as follows:

$$E = hf$$

$$p = k(1/\lambda')$$

Davisson and Germer obtained the wavelength of the electrons' matter waves by using their linear momentum as particles. With those wavelengths, they were then able to calculate predicted values of the angles of scatter, measure the angles of scatter from the experiment, and assess whether they were the same. If you can imagine a drumroll going off right now, please do so, because the result is telling:

They aren't the same! Their predicted angles of scatter were different from the angles of scatter calculated in accordance with Bragg's law. However, the percent difference between their expected and measured values was approximately 1.2%. Maybe the crystalline properties weren't perfectly recorded. Maybe some dust hit the electron gun. Either way, who cares! When the subject of your experiment is to determine if one "thing" behaves like something it would never behave like in an intuitive sense, getting a percent difference of 1.2% doesn't just mean you're on to something; it means you probably found it. Particulate matter (with localized properties) cannot just be thought to, but justified to, demonstrate wave behavior.

What sprung forward from the 1913 experiment's results is what we now refer to as quantum theory. Quantum theory has been through quite a lot in the last century. It's difficult to highlight three to four events that summarize that time period, but one of them I believe unequivocally deserves recognition. In 1961, Clauss Jonsson had an idea that was actually quite simple: take the double slit experiment but send in electrons. He reasoned that if electrons truly behave as waves, then they should logically produce somewhat of an interference pattern. This reasoning shouldn't be impossibly difficult, because every wave known to physicists, when sent through a DSE, produced interference. If electrons were waves, what difference should be expected?

By the time Jonsson performed this hybrid experiment, quantum theory existed. Applications of quantum theory were projected. Quantum theory was even taught to students. It was not as if Jonsson's experiment reaffirmed the hypothesis that electrons were waves; it was the fact that the messy framework of mathematical laws that lumped together to form quantum theory was justified by the double slit experiment. It is not the aim of this book to teach you anything about physics other than that it is possible to think about it in creative contexts. That means I will not (and cannot) teach you what these "mathematical laws" are, mean, and predict for quantum research. If you desire a conceptual understanding of quantum mechanics, that's unfortunately not possible. As Professor Leonard Susskind argues in his The Theoretical Minimum, the mathematics actually makes physics easier to understand. But if you don't understand the mathematics, physics will eat your breakfast, lunch, and dinner.

## THE DOUBLE SLIT EXPERIMENT, AGAIN

Understanding the experiment that Jonsson conducted does not require comprehension of mathematical foundations.

Understanding why Jonsson discovered the results he did does. We'll focus on the former.

Sending electrons into the DSE means you take your plate, you puncture two slits, and you shoot some electrons in to see what happens. Unfortunately, however, you need to configure some properties of the experimental apparatus before anything happens because there's a problem. If you were Jonsson, you have to believe that electrons might just not produce interference, because if you knew they did, you wouldn't be very intelligent to conduct the experiment in the first place. So your underlying assumption is that despite the work of Lester and Germer, there is a possibility that electrons have been particles all along and previous experimenters have been mistaken. You're still testing for interference, meaning you also have to anticipate that it could happen. That means your experimental set-up has to accommodate two possibilities that you can measure:

Electrons are particles, and they should therefore hit the screen randomly. If this is the case, your screen has to give you "marks" on its surface that tell you where particles hit it.

Electrons are waves, and they should produce interference. If this is the case, your screen should simply light up whenever it senses amplitude.

This isn't an either/or situation, which only makes quantum theory more difficult to think about. Fortunately, we can get around it, and here's how. You design a fluorescent marking method that lights up a particular segment on the screen if an electron crashes there, but you also recognize that if an electron is really a wave, it should produce a distribution of intensity as opposed to a spot of intensity. You then turn the electron gun on, observe the pattern on the screen, and make any necessary conclusions. We will ignore Davisson and Germer's finding for a minute and insist that electrons are particles. Consider the potential results of this experiment if our assertion were true.

Do we expect diffraction? Is the electron going to spread out as it passes through the slit? According to a particulate understanding of the electron, this violates intuition, so the logical answer is no. Do we expect interference? Will two electrons constructively and destructively interfere to produce bright and dark bands on the screen? Certainly not, particles don't have amplitude. What do we expect? If you were to use bananas instead of electrons, the bananas would crash through the slits and hit random parts of the screen. If you don't see it, find some bananas, throw them at a wall, and mark where each hits the wall. When you're done, you should have a collection of marks that are located in different locations.[4] Particles should not obey a certain pattern as they travel through this experiment. All that should happen is for the particle, or banana, to travel through any of the two slits, be deflected in some fashion, and hit a random segment on the screen.

Jonsson placed the detector very far away from the plate. The purpose was to isolate the electrons as they entered the double slit experiment to avoid collisions between electrons. Such collisions might produce oddities in your measurements. If you're focusing on electron behavior, it would be ideal to look at what happens to one electron when it is essentially isolated from other sources of disturbance. If you have an electron gun that's placed a great distance away, you can always lower the frequency of emission so that only very few come out; the distance of the gun to plate affirms that only one is coming through at a time.

With one electron at a time, what Jonsson found was a random pattern of dots appearing on the screen. There was no interference at all. But when the experiment was run for many, many hours, after a great number of electrons accumulated on the screen, Jonsson found that the distribution of electrons displayed an interference pattern. This shouldn't be trivial. Electrons are not supposed to act as particles when

---

4    The positions of the marks would depend on how well you aimed.

they enter this experiment individually and produce interference when viewed as a distribution. Though clearly, Jonsson demonstrated that when millions of electrons are sent, the center is the brightest and on the sides are alternating bands of bright and dark.

So, here's what we do know. We know that one random electron, if shot independently of the others, behaved as a particle. It didn't obey the laws of interference. More importantly, we know that there was a moment at which Jonsson shot the gun to send an electron into the experiment, and there was a moment at which the fired electron crashed onto the screen. It went from somewhere to somewhere else; never mind that we don't know anything else, because we know it took a trajectory. Unlike a wave, it had a defined, localized contact point, but many of these little guys wanted to act as a wave.

Let's say you are playing a game with bananas that goes like this. You're standing on top of a building with ten thousand bananas in your hands. One by one, you drop the bananas onto a target on the street at the base of the building. You're told that the target is partitioned into five sections; the middle section is the largest, and any banana has a 90% chance of landing in that section. On opposite sides of the middle section are two smaller sections with a 4% chance of landing a banana in either one. There are two even smaller sections on the very ends of this target with 1% chance of a banana landing in either. This is a statistical game.

When you throw the first banana, you might predict (with accuracy) what will happen. There's a 90% chance the banana ends up in the middle section, but that doesn't mean it can't end up in the 1% section. It might even end up in the 4% section. You have no idea. Probabilities are chances, not restraints. If you let go of one, it could find itself in the middle section. But try letting go of the rest. The second comes in at one of the 4% sections, the third at the other 4% section, the fourth again at the 90% section, the fifth at 1%, etc. As you increase the number of bananas you release, you get more in the middle 90% section, but you also get a number of bananas that make their way to the unpredictable edges. In fact, if you want to get quantitative, releasing 10,000 bananas means 9,000 are ending up in the middle, 800 are going to the 4% sections, and 200 are finding their way to the outer edges. But if you release one, you have no idea what will happen quantitatively, unless you're willing to say that 0.9 bananas lie in the middle and 0.02 are in the edges. That's just statistics. Releasing one at a time doesn't give you the power to make ultra-significant predictions as to where it's going to land, regardless of the probabilities you have to work with. Sometimes, releasing one at a time doesn't even give you the power to make sensible predictions. For example, you can't look someone in the eye with a straight face and say, "I got

0.02 bananas on the outside bands," but you can say, "200 bananas are on the outside bands."

Indeed, if you let go of all ten thousand bananas at one time, it's fairly easy to predict what will happen. Of course, you don't know if exactly 9,000 bananas go to the middle, or if exactly 200 go to the ends, but even if 8,989 bananas come to the middle, that's a fairly good approximation. It's certainly better than releasing one and reporting, "That's 0.02 bananas we've got at the ends."

You're playing the art of probability by using a larger sample. In other words, you're forming a probability distribution of bananas that hit the target in a predictable pattern, as opposed to a single banana hitting the target in an unpredictable location. It just so happens that in the double slit experiment, the most probable locations for electron contact are the bright bands on the screen, and the least probable are the dark bands. Any random electron has a high chance of striking the very center of the screen, but if you only send in one electron, you might not see an electron at the center. Send in ten million, and you get a probability distribution with a large number of electrons at the center and very few where the dark bands are supposed to be.

That probability distribution is defined mathematically as the square of the electron's wave function. In classical mechanics, any physical system you have contains a certain amount of information. You throw a tennis ball straight up in the air, and there is information associated with the ball's trajectory. All of the information in a system can be completely specified by momentum and position measurements. In terms of the physics, you know everything you could know about the tennis ball, at any arbitrary moment, as long as I give you its position and its momentum. In quantum mechanics, however, the laws of classical physics don't apply; I think we've seen enough experimental evidence to make that kind of assertion. Quantum mechanically, you still have information that can be mathematically specified, but no longer do position and momentum measurements make the cut. In fact, one of the more critical postulates of quantum mechanics is that you can't even have momentum and position values at the same moment, which you might recognize as the Heisenberg Uncertainty Principle.

So in order to specify the information contained in the system of double slit experiment/electrons, we need some other mathematical concept. Ideally, we should be able to attach to it the concept of the probability distribution, as discussed in the interference pattern in the double slit experiment. What to do? Define something called a wave function, which is essentially a function that ascribes a value to an electron at all points of space and time. If you put an electron in a box, you have a wave function that assigns a certain value to any position inside the box at any time. Additionally, if you locate an electron at some random time, find the value of the electron's wave function, then square that value, quantum mechanics tells you that this squared result is the probability of finding the electron at said position and time. More concisely, if you want to find an electron at some point in spacetime, you square the electron's wave function at that point in spacetime. You don't get an answer, but you get the probability that the electron is there. The wave function effectively defines everything you need to know about microscopic systems in which electrons behave probabilistically, as they do in the double slit experiment. What that means for the distinction between quantum and classical mechanics should be relatively trivial: there are no such things as deterministic trajectories in quantum mechanics. When you throw a banana at someone, there is a trajectory you can assign to that banana with which you can predict exactly where the banana will be some time from now. In quantum mechanics, not only the time evolution but also the very existence of systems is governed by chance.

## SUPERPOSITION

Here's an interesting question. When the electron is fired from the gun, it has a certain probability of crashing onto any location on the screen, but until it does so, it hasn't decided yet. So where is it?

Yes.

It's known as superposition – a phenomenon that describes how electrons behave until they crash on the screen. Superposition states that from the moment of fire to the moment of crash, the electron is located in a superposition, or collection, of many different possible locations in the source-slit region. It isn't at a defined point, but rather, it is everywhere at the same time, and only when it is forced to decide – meaning when it encounters the screen – does it choose where it ends up on said screen. In addition, the basis of its decision is its wave function, because the square of the wave function is what gives you the probability of finding the electron at any value in space and time. The wave function comes assigned with a set of quantum numbers, which are coordinates that define the location of the electron. The electron, before measurement in this experiment, is superimposed in all possible sets of quantum numbers. When you measure, the wave function observes an interesting property known as collapse – when the wave function "chooses" one set of quantum numbers, and subsequently one location in space and time. What the wave function chooses is what we see, what we see is what the wave function chooses. Anything before the choice is made falls under the expansive and admittedly confusing observation known as superposition.

Therefore, it makes sense for a single electron to not observe interference, because you can't have a probability distribution of wave functions with one electron. You need many of them, and when you experiment with many of them, you invoke the behavior of statistical distributions to give you what is the most probable result – which is precisely in accordance with Jonsson's experiment.

To get a closer look at these postulations, a Japanese technology company by the name of Hitachi Global conducted the DSE experiment in 1993 but with a twist. They knew about wave functions, wave function collapse, and superposition, but they didn't quite understand the full empirical basis of these concepts.[5] To investigate, they intended to find out how electrons "choose" where to land on any particular region on the screen. They built a light detector.

The light detector was placed directly adjacent to one of the two slits and it shined light in front of that slit. The detector's job was very simple: when an electron approached the plate, it had to pass through one of the slits if it were to crash on the screen. As it approached that slit, it would encounter the light beam provided by the detector. That encounter would inevitably produce a measurable collision, and light would be reflected. Meaning, Hitachi Global could observe

---

5   To be fair, nobody really did. At least according to Professor Feynman.

tiny collisions of electron and light beam, providing them an accurate method to assess which slit any given electron was taking to make it to the screen. Why is this helpful? Well, just by looking for an illumination in the region, you know immediately where the electron goes, because it has to go through one of the two slits. Equipped with that knowledge, you have sufficient information to predict where the electron could go. You don't have much information, but you certainly have more than what you had without the detector.

Hitachi found something peculiar. After building the light detector and looking for illuminations, they were able to track where most of the electrons were going in this experiment. For example, if they sent one electron in but did not observe an illumination in front of the plate, they knew that there was no collision between electron and light beam, meaning whichever slit had the light detector shining on it wasn't the slit of choice for that electron. On the contrary, if they found an illumination for the second electron, they knew that the electron took the path of the lighted slit. They were able to detect where electrons were going in this experiment, giving them a new dimension of knowledge that Jonsson never had. However, the result was that the majority of electrons no longer behaved as waves. The majority of electrons hit the screen in random locations, and even after sending in a full distribution of them, they saw many random marks.

Nevertheless, after allowing the experiment to run for many hours, they still obtained an interference pattern that was much weaker than the one Jonsson obtained. It seemed as though just by implementing a method to locate which slit electrons were going through, Hitachi was able to "wash out" the interference pattern and subsequently discourage electrons from behaving as waves. The reason why there was any interference at all was because Hitachi's detector wasn't perfect; there were still some electrons that managed to make it by the plate undetected, meaning those didn't receive the "treatment" that most electrons received from the light source. Here was an interesting observation: electrons that were hit by the light wave behaved as particles but the small number of electrons that escaped the light wave's beating worked together to produce interference.

## AN EXPLANATION?

The conclusion was that when electrons are tampered with, or more generally, when quantum systems are tampered with, they lose their "quantumness." Electrons no longer abide by the probabilistic nature of the wave function if you try to locate them with a light source; you interfere with the quantum delicateness of their nature and prevent them from observing quantum mechanical effects. The result is that they behave as classical particles, or little "things" that hit the screen randomly. On the contrary, electrons that were not tampered with – the electrons that made it by unscathed by the light source – preserve quantum behavior, and thus, give you an interference pattern that is governed by a probability distribution. In short, quantum mechanics is very exclusive, meaning if you try and mess with it, it no longer stays the way it is.

Superimposed wave functions, the heart of the quantum mechanical world, become permanently collapsed wave functions that observe one state and one state only. That's of course what we observe in the classical world. After all, you don't have superimposed bananas. If bananas did observe superposition, it would be possible for a banana to be peeled, unpeeled, peeled half-way through, and peeled 70% through, all at the same time. Not until you look at the banana, or shoot some light at the banana, or touch the banana, does the banana collapse on a wave function and stick with what it collapsed on.

What causes collapse? Macro-scale things. Dr. Ramamurti Shankar describes this in his Fundamentals of Physics, II: everything we deal with on a daily basis is out to get us. There are particles and gases suspended in the air that crash into us as we walk, radiation from devices, and photons of light that tackle our faces everywhere we look. The world might seem serene until you look at the inner workings. Once you add in all of the engineering applications we encounter and employ daily, what you have is a macro-physical world that does not allow quantum mechanical effects to be preserved. Wave functions we deal with are not observing superposition because we have found a way to tamper with their quantum systems to force them to behave as classical objects, just as the light source forced the electrons in the DSE to behave as particles.

Naturally, that means that if you want to observe quantum mechanical effects, you absolutely cannot interfere with the intermediate behavior of a quantum system. Quantum mechanics doesn't want anybody to look at it while it's working, meaning you can fire an electron gun to mark time zero of your experiment, and you can measure the interference at time "end" of your experiment, but in between those two times, you better not let anybody look at what's going on. Unfortunately, that also means that you too cannot look at what's going on.

Think of quantum systems as unpeeled bananas with marbles inside of them. When you look at such a banana, all you see is the yellow skin of the exterior and you can't see the marble inside. The marble, according to quantum theory, is in a superimposed state of multiple possible wave functions. It's not localized in one region inside the banana. All at the same time, it's at the top of the banana, the very bottom, the left, and three centimeters to the right of center. As long as you define the possible locations of the marble, or the states in which the marble may exist in, you have to agree that there is superposition between all of these states. So you don't know where the marble is. However, you can find out.

Peel the thing! As long as you interfere with the isolated, exclusiveness of the banana (the quantum system), you force the marble to sit somewhere. At that point, the party's over.

## COMPUTATIONAL WORK

Recently, there have been breakthrough applications of quantum theory to computational methods. Many public data are encrypted with a specific application of cryptography called prime number factorization (PNF). Take two very large prime numbers, for example, 26,863 and 37,579, and find their product. That much is simple. If you try to work backwards, in other words, take some random, huge integer and attempt to come up with two prime numbers whose product is that integer, you probably wouldn't be able to do it. Also, you would have no reason to feel bad about it, because the world's smartest mathematicians haven't been able to do it either.

Computer algorithms have been designed to tackle PNF and they have worked. Currently, there are algorithms for factoring integers but there are some ramifications. The first is that nearly one thousand computers, wired in parallel, provide the minimal computational power for such a task. The second is that factored integers have only been two-thousand digits long, which might seem large, but is relatively small compared to the level of security associated with modern encryption software. Additionally, when one-thousand computers work together to factor the integer, it takes about two years. While computers have done marvelous things for us in industry and technology, they have not been able to supersede the bounds of number theory and information science, which require much more powerful computational grids. Especially for a task as difficult as PNF, we're going to need a bigger computer.

In 1994, a mathematics professor at MIT by the name of Peter Shor theorized the potential for a computer to correctly factor an integer to its primes efficiently. But Shor's Algorithm, although a theoretical wonder, was not feasible at the time. For it to work, Shor needed a computational technology with a particular attribute: it had to have what's called qubits.

In a classical computer, information is stored in a type of language called binary code, which is a number system that utilizes two values: 0 and 1, high and low, or on and off, which is what you might find on computer science wallpapers. Information science moves too quickly to program the computer to understand our language, so data is compressed into little packages. Software is designed to allow the processing and execution of tasks depending on what the packages read. If you want to tell your computer to lower the brightness, you don't just say, "ok computer, 0-1-1-0-1-1," because that's just binary gibberish.[6] You press a button on the keyboard, and you wait for the computer architecture to allot less voltage to the lighting that makes the display visible. That process involves the transmission of instruction to relevant components of the computer, ranging from the CPU to the screen, both of which rely on binary code.

The 0's and 1's of binary are called bits, and their states are, hence the name, binary. There's no rebellious teenager bit that wants to be 0.125 bits or half a bit. Every package of information in a classical computer is represented as one of these two values, meaning once some letter or number is assigned a string of bits, there's no going back.

This should seem like the opposite to many topics I discussed previously. Unlike a classical computer, quantum systems are not binary; inherently, a quantum state is in a superimposed state of multiple different wave functions and if you ever want to find out which one it's in, it collapses on one of them for you. That's ultimately what you see in real life, even though you know its wave function was once much more than that.

---

6   Actually, it reads "27" in our everyday number system, but you get the point.

Over the last decade, computer scientists have found ways to invoke the power of quantum systems to design qubits. Qubits are quantum particles that form the basis of information storage in quantum computers. Essentially, they are tiny information-carrying particles that are superimposed in the two states of 0 and 1. Only when you measure – or tamper – with the qubit does it collapse on one value, thus giving you a classical bit.

Now you might tell me that's pointless, because for example, if you have four classical bits that form some binary system, each one can be in two states, meaning 2×2×2×2 gives you the number of states you can have, which is 16. However, if you have four qubits that form another system, the maximum number of states is still 16, because after measuring, each qubit still has to have a value of 1 or 0!

Except there's a catch. For classical computers, there are most certainly 16 possible states, but the system will, initially and forever remain settled on only one of those states. If the string of bits for a four-bit set is 1,0,0,0, then 1,0,0,0 will be the string forever. But superposition allows a quantum configuration to exist in all 16 states at the same moment in time, so it does have the same ultimate potential of a classical system, but it fully maximizes its potential every single time. In a quantum system, the qubits can theoretically be in states (1,0,0,0), (0,1,0,0), (0,0,1,0), and the list continues. Before you measure the qubit collection, it will occupy all possible 16 states. This means it has the information-storage power of 16 different classical bit systems that each occupy a different state. That is the beauty of the wave function and quantum mechanics in general. Multiple pieces of information can be stored in a system in many different states of information storage. Superposition gives the quantum computer the ability to input more information than has ever been seen in classical computation. It gives tremendous algorithmic power in terms of efficiency.

Researchers at MIT, in March of 2016, used a quantum system with five qubits to successfully factor the number 15 down to its primes, 3 and 5. Granted, 15 is a very simple number in that there are only two digits and most people who have graduated the third grade can tell you that two of its factors are 3 and 5. But the real beauty of this finding is that Shor's algorithm worked on its intended platform (the quantum computer!) and qubits did what MIT told them to do. Never before has that happened. There is a tremendous potential for quantum computers to tackle bigger and better numbers, because companies such as IBM have found ways to implement 50-qubit processors in a quantum computer. In March of 2018, Google topped that with a 72-qubit system. By the time you read this, there will probably be a start-up or technology company that will have exceeded 72 qubits. With more qubits and subsequently more effective use of quantum behavior, researchers are looking toward applications in cancer genomics, image resolution techniques, security encryption, blockchain models, financial analytics, and a variety of other fields.

Perhaps the most prevalent challenge is the fact that in order for a qubit system to observe superposition and subsequently enhanced storage of data, the quantum delicateness of that system has to be preserved in its entirety. You can't mess with the system, you can't look at it, and you can't even let qubits touch each other. All of those situations would lead to wave function collapse and loss of quantum behavior. You can then imagine how precise experimental set ups and how carefully wired a quantum computer needs to be in order to run two qubits, much less 72. Another problem is that most qubits used need to be preserved at temperatures very close to 0 Kelvin, which is about -270 degrees Celsius. That's pretty cold!

The fundamental postulates of quantum theory, at least the ones outlined in this chapter, are as follows. Electrons individually are particles, but their position in space is defined by a complex wave function whose square gives you the probability of finding the electron at any given location and time. In the double slit experiment, the wave function allows you to calculate the probability of finding any given electron at a point on the illumination screen, but you know statistics doesn't work that easily when you only have one electron. You therefore send in a distribution of electrons, just as you throw off a whole bunch of bananas off the building's roof, and you see the spatial probability distribution on the screen as the interference pattern. Therefore, electrons observe wave behavior.

That only applies in isolated quantum systems, when there is nothing to tamper with the delicateness of electron behavior, such as a light source. When such conditions for isolation are met, electrons exist in superpositions of many possible states. The very act of measuring the quantum system tampers with that delicateness and forces the electron to collapse on a wave function. When that happens, it reaches a classical, permanent state.

In classical mechanics, the state is established, and we need only find out what it is by breaking out some detectors. In quantum mechanics, things are different. The state is laughing at you, playing dice on a table, and watching you fumble around with whatever detector you brought in your bag. Even without the mathematics, quantum packs a punch.

# PART 2

## ALBERT EINSTEIN

# 4

# BANANAS ON THE RUN

—

Special relativity is different from quantum mechanics in that all experiments conducted are not on microcosmic realms of observation; special relativity is similar in that experiments conducted are still on radical realms of observation. This means that both ideas are introduced when physical observations exceed the scope of our classical world, or what we interact with on a daily basis. In the case of quantum mechanics, electrons are simply too small and too delicate for us to observe and interact with while they observe quantum-like behavior. Quantum mechanics describes the "too small." The parallel for special relativity is "too fast."

Between 1905 and 2018, physicists have been performing experiments to verify the theory of special relativity. To date, every experiment that has been thrown at it has given data that support the postulations made by its author, Albert Einstein. Why this is important is because when special relativity was written down, it screamed contention with classical ways of understanding the motion of objects in a relativistic sense. It was widely regarded for its simultaneous mathematical simplicity and intuitive complexity. The theory was neglected by many, but was eventually hailed as the single most important paper of the turn of the twentieth century. That's all because Einstein was bold enough to rewrite a few laws of physics that he didn't agree with. It's not the same thing as "failing to support a hypothesis." It's moving to the next line and writing a hypothesis that better reflects observations made in experiments. The reason why Einstein's "next line hypothesis" is so popularly known today is because experiments have been done that support it tremendously, despite how much it might not make sense. That's why it's important.

## COORDINATE SYSTEMS

If you take a banana and throw it at a wall, you could not simply say, "the banana is moving" to a physicist. In conversation, you can most certainly say that, because we generally are not as smart as physicists are in a quantitative sense. In terms of the physics, you have to specify the relative to what part about its motion. We don't often include this specification in describing things in everyday life. For example, if you tell me, "I drove this morning at ninety miles an hour," I assume you meant, "I drove this morning at ninety miles an hour, relative to the stationary highway," because that is a conversational convention that we all agree to understand. In physics, however, that convention becomes a critical component of what's called relativistic motion, and we can no longer afford to leave it out. Sure, you threw a banana and it's now moving. But relative to what? Relative to an apple flying ten miles an hour next to it? Relative to a plane flying four hundred miles an hour above the room? You have to deliver the context of other objects, whether they're stationary or moving. Only then can a quantitative description of an object's trajectory be provided.

So how do we do that? Relative motion boils down to the concept of frames of reference. Most people are familiar with the Cartesian coordinate system in mathematics. There are three perpendicular, or orthogonal, axes labeled x, y, and z.

The intersection between the three axes is referred to as the origin, and you can pick the origin up and put it anywhere you want, placing yourself in an entirely different coordinate system. If you put an origin at the corner of a table, the x axis might go off to one edge of the table. The z might go off toward the other edge, orthogonal to the x. The y goes straight up, off the table, at the ceiling. But that's not the only coordinate system on the table. You can equivalently locate a coordinate system at any other point on the table, on the edge of the table, on the bottom of the table, or even above the table's surface. Those are all viable coordinate systems.

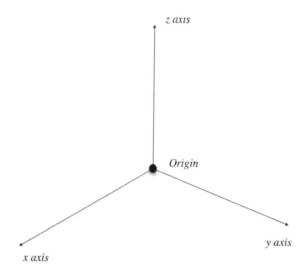

As long as you establish a coordinate system where you are, you can measure how you move around. For example, if the origin is at the point where the banana leaves your hand, you can measure how far the banana travels in the x, y, and z directions until it hits the wall. When it does, you can make a new position measurement for the banana.

As much information as these three coordinates might provide, there is more information that seems obvious but is not. When did you throw the banana? When did it land? Cartesian coordinate systems don't tell you when things happen but we can nevertheless make "when" measurements in our experiments. With our clocks, we can figure out when the ice cream truck comes down the street or when the banana hits the wall. In some way, we have to find a unification of Cartesian coordinates and time measurements.

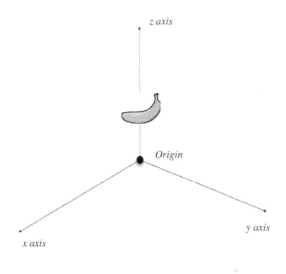

To do so, we can utilize a concept in physics that was developed in the 1700s known as Newtonian relativity. Although Isaac Newton got the name, he didn't develop it himself. Much of the inspiration and intuitive progress came from an astronomer and philosopher by the name of Galileo Galilei.

The problem facing Galileo was very similar to the problem facing us as we try to describe the relative motion of the banana moving at the wall. Galileo's solution, however, was quite simple: you don't need to care about a unification of position and time measurements. According to him, space and time were two different areas of observation in our universe and it failed intuition for them to be intertwined. If you throw your banana at the wall five meters away, and you are at the origin of your coordinate system, the banana ends up at x = 5. As long as you have your clock in the other hand, you measure that the banana hits the wall after two seconds from the moment you threw it. So the banana now receives spatiotemporal coordinates: x = 5 meters, t = 2 seconds. If I'm standing ten meters behind you, at my own origin, the banana hits the wall at my x = 15 meters, so there is the obvious distinction in our position measurements. Except, as long as I start my clock at the same time that you start yours, we both get t = 2 seconds. My banana would just have to move faster.

Not every person needs to walk around with a clock wherever they go, because time is absolute. Every clock in the universe runs at the same pace, reads the same time, and will always read the same time no matter where you go. That was the conclusion of Galileo and subsequently of Newtonian Relativity: position is a relative measurement, meaning it matters where you stand, but time is absolute.

## BRIEF EXAMPLES

Let's imagine a random coordinate system in a park. You go into this park, sit on a bench, and watch two birds flying north. Bird A is flying at 20 meters per second (m/s), and bird B is flying next to it at the same speed. But relative to what?

Relative to the park's stationary coordinate system. The park can be considered a static background space for birds to fly in and people to sit on benches. The park is not getting up and walking to buy a coffee somewhere, which means the park's coordinate system is not moving. Also, because the bench on which you are sitting is part of the stationary park, you are also stationary. Therefore, the birds are also flying at 20 m/s relative to you, a stationary observer.

However, let's say a guy on a motorcycle comes rolling through this park, in the direction in which these birds are traveling, at 80 m/s. Motorcycle guy is a different observer than are you: motorcycle guy is moving and you're not. If motorcycle guy looks up to see the birds flying, he's going to measure their speed to be different than what you measured it to be.

If objects move relative to each other, they can make measurements about what is moving around them, but those measurements will not be normalized to a standard coordinate system. They are dependent on the object's own motion.

When you drive on a highway and look at a car next to your car that is driving at the exact same speed, the car doesn't look like it's moving. You can look at the speedometer and measure that you both are moving at 80 mph, but if your friend is sitting in the backseat, watching the adjacent car, the friend wouldn't think the other car is moving. If you slow down your car a bit, perhaps to 60 mph, and the other car stays at 80, it's going to look as if the car is crawling slowly forward, at 80−60 = 20 mph. As you slow down further, the other car looks like it's getting faster, if you speed up, it gets slower. If you stop your car, the car zips by at 80 mph and the experiment is over.

In the park example, you are stationary relative to the birds and you therefore measure their speed to be 20 m/s. However, motorcycle guy is moving so fast in the same direction that he doesn't measure the birds to be moving in the same direction. All he has to be traveling at is 20 m/s for him to think that the birds are stationary relative to him, because that's when they have the same speed. As soon as he goes faster, it looks to the birds as though motorcycle guy is moving forward, and it looks, to motorcycle guy, that the birds are moving backward. You can measure the birds to be going at 20 m/s in one direction, but the birds can equally say that you are the one moving, just in the opposite direction. The birds think motorcycle guy is going at 80−20 = 60 m/s, but motorcycle guy could easily say that he is stationary, and the birds are really the ones moving. They're just moving the opposite way. There are all kinds of statements you can make about the diversity of measurements in this coordinate system.

You can also have relative motion of coordinate systems themselves. Let us say, for example, you take a friend to a highway and get into a car. Your friend stays on the side of the road. You can define your origin to be where you are, your friend can define another origin where she is, and you both have two coordinate systems. Better yet, if your car starts moving, you don't have to be an object moving in a coordinate system; you are an object moving with the coordinate system, because the coordinate system is itself moving. If you drive past your friend at 60 m/s, for example, what can you say about the relative motion between you two? You know your friend will say, "You're moving at 60 up the road," but you don't have to agree with her if you don't want to. You could say, "No, I'm at rest, you are moving at 60 in the opposite direction." As long as you both understand that you will measure in different directions, you're both right. This might sound very similar to the park example, but it's not: with the park, there is a coordinate system in which objects are moving. This time, coordinate systems are moving themselves.

Generally, if we locate ourselves in coordinate systems that are moving, we can detect that our coordinate systems are moving. If you are driving on a highway, it's hard not to know you are moving. You see the "WELCOME TO NEW JERSEY" sign, you see the mile markers, or you see other cars going by. If you are a bird flying through the park, you know you are moving. Maybe you see some person on a bench down below and it looks as though the bench is moving backwards. Maybe you see a plane up above that seems to be flying forwards. The very notion of watching something that is moving suggests to you that you are moving, because as long as the "something" claims itself to be stationary, it is a physical requirement for you to be in motion.

However, this is not always the case, and Galileo was the first physicist to describe why and how. Galileo said that there are ways to be blinded. Under certain physical conditions, it is impossible to exist in a moving coordinate system and conduct an experiment that informs you of your coordinate system's motion.

What are such physical conditions? There are two: your coordinate system must not be accelerating and no net forces can act on your coordinate system. If you know a thing or two about Newtonian mechanics, you know that these aren't really separate conditions because one necessarily implies the other. Forces can act on objects to get them to start accelerating. Acceleration is defined as a change in an object's velocity, and velocity is simply speed with a direction attached to it; saying you move at 20 mph on a road is a speed measurement.

It is different from saying you move at 20 mph northeast, which is a velocity measurement. If a force acts on an object, it produces acceleration, meaning the object's velocity changes. How much it changes depends on the object's resistance to being accelerated, which is properly known as the object's mass. They are all related in this equality:

$$F = ma.$$

Force is equivalent to mass multiplied by acceleration. If you give a banana a force of 2 newtons (unit of force), and the banana weighs one kilogram, the banana accelerates at 2 meters per second squared.

I'll remind you of the condition Galileo put forth before we go any further. As long as the coordinate system you are in is not accelerating and you have no external information that might suggest you are moving, you will believe that you are stationary. External information quite literally means information that "ruins the joke." When we travel on a highway and we see the "WELCOME TO NEW JERSEY SIGN" zip behind us, that's external information. Galileo said observers must be blind to such information in order to truly be unaware of their movement.

You have to be in a non-accelerating coordinate system, or as Newton later called it, an inertial coordinate system. Inertial, or the property associated with zero acceleration, is defined in two scenarios: the first is if the coordinate system is not moving at all. This should be easy to digest; if you're sitting in a coordinate system that's not moving, you're not going to think you are moving if you have no external indication that might state otherwise. The second scenario is less obvious: recall that acceleration mathematically translates to a change in a velocity measurement over some time interval, meaning the bigger the change in velocity, the greater the acceleration, but if there is no change in velocity, there is no defined acceleration, meaning coordinate systems that maintain a fixed speed are also inertial. This brings us to the critical assumption: somehow, in some way, if I am in a coordinate system that moves at a fixed speed, you are in your own coordinate system that is stationary, and we don't have information from the outside world, we can both convince ourselves that we are in coordinate systems that are moving at fixed speeds.

Similarly, we can also convince ourselves that we are in coordinate systems that are not moving at all. This is, of course, because our two coordinate systems are inertial. We really have no idea which coordinate system we are sitting in: perhaps one that moves at a constant speed, perhaps one that moves not at all. More technically, if we both conduct random experiments in our separate coordinate systems that are the same experiment, we get the same results. Non-inertial to both of us can take two possible definitions that we cannot formally decide on without external information. If I drop a banana in my coordinate system and watch it fall with a certain acceleration, and you drop a banana in your coordinate system, you will observe it falling at the same acceleration. It doesn't matter that your coordinate system, or my own, might be moving at fixed speeds or not moving at all. We both observe the same laws of physics.

Admittedly, Galileo didn't write this principle down with bananas. He said that if you were to sit below the deck on a ship that was moving at a fixed speed, on a perfectly calm, smooth sea, and you could not see outside of the room you were sitting in, you would not be able to tell if the ship was moving or if it was stationary. As an analogy, suppose you are reading this book in a room. The room is moving at a fixed speed. The air in which the room is moving through is perfectly calm, meaning no force can alter the room's acceleration. You also can't look outside of your room, meaning you can't read the "WELCOME TO NEW JERSEY"-style sign to figure this experiment out beforehand. Galileo said that there is no way for you to know that the room is moving because on the inside, it looks identical to the coordinate system of a room that is stationary.

Now consider this: let's say you are moving in Galileo's ship with a banana that is placed on the floor and there is nothing else inside the ship with you two. Just you and the banana. You have to agree that the banana and you yourself will simply assume the motion of the ship's coordinate system, just as you can be stationary in the car on a highway. So, recall Galileo's postulation: you have no way of knowing that this ship is moving. Putting a banana on the floor doesn't give you any more information about the ship's movement other than what you already had beforehand.

Let's say you have been buckled down to the floor of the ship and the banana is placed on the floor in front of you, but the banana has not been buckled down. All of a sudden, the ship stops moving. When the ship stops, the banana flies forward. Just because the banana assumed the movement of the non-accelerating ship before doesn't mean the banana will assume the movement of the ship now. In addition, you have experience with this kind of thing: when you are on a highway and the car halts instantaneously, you fly forward. That's why the airbag is there! Similarly, if the ship were to abruptly stop, the banana would shoot forward.

Because the banana shoots forward after the stop, you can immediately deduce that the ship was once moving and it abruptly came to a stop. The explanation is that the banana no longer kept the inertial coordinate system of the ship; when the ship came to a stop, the banana stayed in its own coordinate system. It went from a non-accelerating coordinate system to an accelerating coordinate system: inertial to non-inertial.

If you're careful, you'll notice that this agrees exactly with Galileo's theory: two inertial coordinate systems observe the same laws of physics, because they are identical to whichever observers you put in them. However, two coordinate systems that are non- inertial, meaning one accelerates and the other doesn't, observe different laws of physics. As long as the banana's coordinate system starts to accelerate, it's perfectly fine for it to move as it does. If you try to use Newton's laws on the banana (meaning measure its acceleration and calculate the force on it), it won't work; if nothing else is in the ship with you and the banana, there's nothing that can produce a force on the banana! Nevertheless, this does not fit Newton's definition, because in order for the banana to change its velocity (assume an acceleration), it had to have been acted on by a force. Indeed, non- accelerating coordinate systems do not have to observe the same laws of Newtonian physics as those observed by accelerating coordinate systems.

This essentially sums up the Newtonian theory of relativity. Relative motion of objects in coordinate systems works by adding or subtracting velocities appropriately. If I'm flying a plane at 90 m/s, and you are flying next to me at 80 m/s, it looks to you as though I am flying at 10, and it looks to me as if you are flying at 10 in the opposite direction. In addition, if we are in different coordinate systems that are not accelerating, and we can't see external indicators of our motion, we won't know if we are moving at constant velocities or if we are stationary. We both observe the same laws of Newtonian mechanics, meaning F = ma.

## WHERE LIGHT CAUSES TROUBLE

Pragmatically, meaning the scale on which we define everyday objects, Newtonian relativity works. This is why you don't think the car adjacent to you on the highway, which is moving at the same speed, is moving. However, once physics becomes a bit more extreme, Newtonian relativity fails. We'll see this as a common theme in the history of physics: whenever someone has introduced an area of physics that is radically different from intermediate levels, "intermediate level physics" no longer works for all situations.

It turns out, what isn't so intermediate about physics is light. You know already that James Clerk Maxwell, after experimental support from Thomas Young, concluded that light is an electromagnetic wave. The wave velocity of light, as it was calculated, is about 300 billion meters per second. Before we proceed, you might want to take a minute to admire this observation: physicists managed to calculate the speed of light to be 300 billion meters per second. This means if you sent light to circle around the earth, it would take less than

0.01 seconds to complete its circumference. Do recognize that this speed is very much not intermediate: normally, we travel around one ten-billionth of this speed on highways.

If you remember the definition of a wave, or a moving "thing" in some medium, you know that in order for light to be a wave, it also has to travel in a medium. Sound waves, for example, can travel through water via SONAR transmission. Earthquakes are technically waves and their medium is the earth. The question for light is not "is there a medium?", because that's too simple and frankly too dangerous. Maxwell and Young had better hope there was a medium, because if not, the legitimacy of their findings might have been compromised. The question(s) were: where is the medium, what does the medium look like, can it be seen, and what is it made of?

Because nobody knew how to answer any of these questions, most physicists of Maxwell's time believed that the medium was simply an infinitely large, three-dimensional fluid that permeated all of space. It was invisible, undetectable, and nonreactive with other elements in the atmosphere. It was just ether, carrying everything in the universe.

In general, whenever a wave travels through a medium, its speed increases if the medium is denser. This is why sound waves travel faster in water than they do in air, because liquids are generally denser than gases. We've already established that light moves rapidly fast, so fast that there is not a single object that was known to compete with it. In fact, 300 billion meters per second was so famous that it received its own variable, c. The assumption was that light had to move through a very dense medium, if it was to propagate at c, resulting in the necessary condition that the ether would have to be incredibly dense. However, if light moved through a very dense "ether," and every other celestial body was also contained in the same medium, how do you explain Earth orbiting the Sun at 67,000 miles per hour? Try doing a jumping jack in a gymnasium; not too bad! But fill the gymnasium with molasses and doing jumping jacks becomes a whole lot more difficult, because the molasses gymnasium doesn't want you to move around. Similarly, Earth could not move that quickly if there was that dense a medium that surrounded it. Dismissing the ether argument meant a revisit to the critical question of was light a wave in the first place?

In 1887, two physicists by the names of Albert Michelson and Edward Morley set out to discover this ether. They conducted an experiment known as the Michelson-Morley experiment and the purpose was simple: calculate Earth's speed relative to the ether.

The ether was assumed to be stationary, which meant the Earth, as it crashed through space around its orbit of the sun, had to have some measurable velocity relative to the stationary ether. You might imagine that this was a rather hefty task. It's not easy to simply measure Earth's speed relative to an invisible medium that nobody knew anything about except for the fact that it was there. However, Michelson and Morley managed to pull off an experiment and it is widely regarded as one of the most phenomenal experiments in the history of physics. It was beautifully intricate, precise, but unfortunately, very much unsuccessful.

Imagine you are a runner who always runs at a fixed speed no matter where you are. You've been running at this speed all your life, but in some peculiar way, you have never known what this speed is. One day, you decide to find out, but not using the conventional timer method. Let's say I give you a special speed detector that you can use to measure your speed. The special part about the detector is that it only works in windy areas, meaning you can't simply travel to an indoor facility, start running, and get your speed. You have to use winds. So now what?

Understand that this is a bit of a problem. Because this detector only works when wind blows, any speed you measure is not your true running speed because the wind would have affected the reading. In order to use this detector to measure your fixed speed, you have to find an area where there is a fixed wind. It should travel in one direction, at one speed, always. I know there aren't too many places in the world where such conditions are met, but for our purposes, let's say such a place does exist. You go to this area and find a perfectly flat field, one that won't alter your running form or speed.

You're told that the wind speed in this area is always 3 mph north, and you're also shown which direction north is. You now start running in that direction, such that you are running with the wind. If you've ever done this before, you know how good this is: a velocity booster! Without the wind, you would run relative to the still field, but because there is wind, you run relative to the windspeed. This means if the windspeed is 3 mph, whatever speed you measure on the detector will be your original speed added to 3 mph. If you get 9 mph on the detector, you can do some simple subtraction to calculate that your normal speed, in no wind, is 6 mph.

If you try running in the opposite direction and measure your speed, perhaps you could check your answer. Let's say you run south at your constant running speed and let the wind do its work. This time, the wind is crashing against you, meaning you will be slowed by the wind. If we were right in our calculation of 6 mph, the speed you measure on the detector should come to 6−3 = 3 mph. You can do this experiment yourself if you want to, but I'll save you some time by stating it: you get 3 mph against the wind and 9 mph with the wind, meaning your speed is always 6 mph.

Michelson and Morley had the same logic in their own experiment. The idea was that the ether was the atmospheric environment of this perfectly flat, still field. As the Earth, or the runner, moved through space, it crashed against the ether and generated an ether wind. For example, on a windless day, you can hop into your Ferrari and drive down the highway, feeling the wind on your face. Similarly, the ether was still, but because Earth moved through so rapidly, there had to be an ether wind and it had to be constant. If they could measure the speed of the Earth with and against the wind, they could do their subtractions and find out the speed of the Earth without the wind. This is similar to how you have a fixed running speed, but because we knew the wind speed was constant in the field, we could find your running speed with some simple calculations.

Rather than runners, they chose light. If they could send a light beam moving in one direction and send another light beam in the opposite direction, they could do some analogous calculations: because light moves at c with no interruptions, the light that moves in the direction of the ether wind will travel at a greater speed, say c+x, and the light that moves in the opposite direction will travel at a lesser speed, say c−x. As a reminder, c is the speed of light (300 billion m/s) and x is the postulated speed of the ether wind. I won't talk to you about the experiment set-up, because that takes away from the general point: this experiment failed.

The light beam that was supposed to travel with the ether wind was supposed to move at (c+x), but it was measured to travel at c. The other light beam was supposed to move at (c–x), but it also traveled at c. Interestingly, the speed of the ether wind was always coming out to zero, but even more interesting was the consistency of the speed of light. This is really what intrigued people more: light never changed its speed in this experiment. Furthermore, throughout the 19th century, there was an influx of experiments in which physicists attempted to force light to move at a different velocity in space but attempts perpetually failed. Even today, attempts perpetually fail. Light always moves at c.

I think you know what's coming next. Let's say you get into a car and travel at 20 m/s down a flat road. In about two seconds, you throw a banana from the car and your banana throwing speed is also always constant at 1 m/s. However, because you move at 20 m/s, the banana is also moving at 20 m/s, meaning the banana travels at 21 m/s relative to the stationary, flat road.

Now, let's say you go and buy a device that shoots out light. You take the device into the car with you, again while traveling at 20 m/s. When the car drives by an observer on the side of the road, the device is already moving at 20 m/s, meaning the light that leaves doesn't just move at c; it moves at c+20. This is according to Newtonian relativity, which instructs us to add and subtract velocities whenever it becomes necessary. You see the light leave at c, the observer sees the light leave at c+20, and everybody is happy. Except for Michelson and Morley. Light is supposed to move at c, always! How can it be that some guy on the street saw it going 20 meters per second faster than c? The answer is that the guy is wrong; he's not supposed to measure c+20. He measures c, and so do you.

Light breaks Newtonian relativity. It doesn't care which coordinate system it moves in, or how coordinate systems move relative to each other. It doesn't care about the ether, about the ether's wind, or about anything physicists can throw it. There were two penultimate decisions that could be taken: we're wrong about light, or we're wrong about Newtonian relativity. Both options require deep reformations to what the current understanding of relative motion was in the 18th century.

# 5

# RELATIVISTIC BANANA LAUNCHERS

---

In 1905, Albert Einstein published On the Electrodynamics
of Moving Bodies, a paper that introduced special relativity.
Although not cordially accepted at first, academic professionals eventually came to recognize that Einstein's theory was
revolutionary, crazy, and questionable. However, the mathematics of Einstein's arguments were elegant and concise. If
one were to consider the number of times special relativity
has prevailed in experiments conducted to test its legitimacy,
the theory looks pretty convincing.

Theoretical statements are not synonymous with factual
statements. That difference becomes critical in physics. The
theory of special relativity was written down by a brilliant

man who saw the world through a different lens and was not afraid to share the architecture of his vision. Neither did Einstein expect nor receive immediate affirmation from every other physicist in regard to the veracity of what he proposed; actually, special relativity was at first considered a step too far in the direction of foolish. Only when special relativity "passed" test experiments of the 20th and 21st centuries, primarily in the field of particle physics and astronomy, did the skepticism dispel significantly.

Einstein admitted to not knowing many things for sure, but one thing he absolutely knew for sure was the constancy of the speed of light. He reasoned that c is universal, absolute, and resistant to whatever obstacle you could hurl at it. If I am sitting in a coordinate system that is moving relative to your stationary coordinate system, I could shoot a banana out of my banana launcher, and you would measure a different banana speed than I would measure. However, if I instead carried a light emitter, we would both measure the same speed. A more sophisticated explanation I read about in Dr. Ramamurti Shankar's lectures on physics follows.

## THE TRAIN PLATFORM

Imagine a man standing still on a train platform. All of a sudden, two lightning bolts strike on either side of the man such that he is exactly in between the two points of contact when light strikes ground. Let's say that distance is one meter. So far, you should understand that there's a guy on a train platform and one meter away from this guy's center, on both of his sides, lightning strikes on the ground. To his left one meter, there's a big mark on the ground from the lightning and to his right one meter, there's the same mark.

Now if the man was forced to look straight ahead, the only way he would see the two lights is if the light traveled from the points of contact on the ground to his eyes. If the light crashed at equal distances away from this guy, and if light moves at the same speed, then it would take the same time for both light beams to reach his eyes. He'll therefore measure the two lightning strikes to be simultaneous events, meaning they happen at the same time.

Consider a train that is coming in from the man's right side. In the train is a woman who is standing still and looking out of the train's center window. When the train passes the man, the woman will see the lightning strikes. Because she is coming in from the guy's right side, she is moving toward the lightning strike that happens on the guy's left. The same principle applies here: in order for this woman to measure when these lightning strikes happen, she has to see them. Let's say she's moving at a steady 20 m/s. When she approaches the man, she is moving towards the left lightning strike and away from the right lightning strike. Therefore, she will logically see the left lightning first, because she is moving towards the left lightning.

If you don't understand this, think of this analogy. You are standing still in a field. Suddenly, somebody puts two cheetahs that run at the same speed on either side of you. One is twenty meters to your right and the other, twenty meters to your left. Assume the cheetahs are hungry. You are going to be eaten by both cheetahs at the same time because the cheetahs move at the same speed. However, let's say that one of these cheetahs is a friendly cheetah and it's going to move toward you and ask you to jump on its back. Once you get on its back, the other cheetah can't eat you anymore. But, there's a problem: cheetahs are fast. If you really wanted to make it to the friendly cheetah before being eaten, you'd have to move really fast. If you walk at 0.000004 mph, which is the same thing as traveling at 20 m/s next to light, the other cheetah is going to eat you for breakfast, lunch, and dinner. However, if you move comparable to the speed of the cheetahs, you can outrun the mean cheetah and get to the happy cheetah in time to not get eaten. Similarly, the woman, if she wants to see the left lightning strike first, must be moving at a very fast speed because otherwise, the light from the right catches up. If she comes in at 20 m/s, she will not see the left lightning strike first, but if her incoming velocity is 0.75c, the story is different.

So what does that mean? If the woman is moving at a sufficient velocity, she sees both lightning strikes, but they happen at different times. What was simultaneous for the man is not simultaneous for the woman. This itself is a radical departure from Newtonian Relativity, in which all events are simultaneous in all coordinate systems. The idea of a master clock, one that ticks at the same rate for all observers, is not correct according to Einstein. According to him, if a master clock did exist, the man couldn't observe the two lightning strikes to be simultaneous while the woman observed the two strikes to not be simultaneous. In order for both of them to say, "this lightning strike happened at time x, this strike happened at time y," and have the same values for x and y, time would be absolute. Yet, that's not what they say. Time is relative.

The question of "how so relative?" is also answered in the train analogy. Because the woman observes a time difference between the lightning strikes, she has a time gap in her clock. The man says "these two strikes happened at the same time, because I measured the time difference between them to be zero," whereas the woman says, "my clock must have been moving a bit slower, because I measured the time difference between strikes to be nonzero." The woman's clock has a lag. If the woman did not move at a velocity comparable to that of the speed of light, she would have seen both lightning strikes at the same time. It's as if you got into the cheetah field and ran at a steady 0.00004 mph towards one cheetah; before you know it, the hungry cheetah is asking for the dessert menu. You have to be moving at a fast enough speed, and because we established that the woman is moving at 0.75c, you can make this statement:

Because the woman is moving at a respectable velocity compared to c, she can measure a time difference between events that are supposed to be simultaneous in a stationary coordinate system. The woman's clock passes slower as she approaches the speed of light.

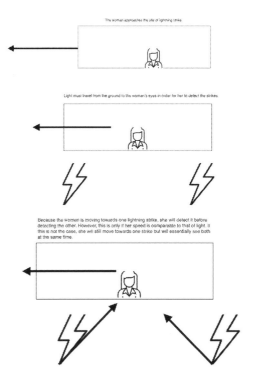

Imagine now there are two pigeons, pigeon A and pigeon B, who both want to measure the length of a pond. When you measure the length of something, you're not simply taking a meter stick and observing how far it extends in one direction. To obtain the true length of something, you have to mark two points at the same time and measure the distance between those points. So, the pigeons somehow have to find a way to mark both ends of the lake simultaneously. You know where this is going. If you have to mark two points at the same time, and time is relative, does that mean the distance between those two points will be relative?

Well, let us say one bird chickens out early, and it wants to simply sit on the side and measure the pond's length from there. It can theoretically measure the pond by making simultaneous measurements of either end's positions, which means it will obtain the true value of the pond's length. It doesn't matter what its friend gets, because it knows that the length measurement it obtained is the true value. Neverthless, the other bird still wants to fly over the pond, so it does so at an extremely fast speed (close to c). What does it see?

We know from time dilation that the flying bird's clock will tick slower as it makes this measurement, since it is in a rapidly-moving CS. Since its clock is slower, the time difference it finds in moving from the pond's base to the other end will be larger, which means the distance it measures must be smaller as well (assume these kinds of birds always move at the same speed). The length of the pond, according to the moving bird, is smaller than the length as measured by its friend. There is an apparent length contraction, which is a direct result of time dilation, which is a direct result of the relativity of simultaneity. So distance and time both get smaller if the speed is sufficiently increased.

Although Einstein is credited with the inception of the relative simultaneity idea, both length contraction and time dilation were actually crafted by a physicist by the name of Hendrik Lorentz. They are together referred to as the Lorentz Transformations.[7]

Somehow, time dilation affects length measurements. In some peculiar fashion, the when and the where of physics are intertwined. How can this be? Galileo and Newton both said that the universe was composed of infinitely many possible coordinate systems, with one master clock that ticked away for everybody. However, it appears as though the flying bird has a slower clock than its stationary friend does. The man in the train sees the same two strikes happening at different times, even though the woman sees them happening at the same time. Their clocks don't move at the same time. What is going on?

---

7 Along with the two mentioned, there is another Lorentz Transformation not discussed here that pertains to an object's mass.

## SPACETIME: A COUNTERINTUITIVE PAIRING

Einstein embedded this in the idea of spacetime. The universe is not separately composed of spatial coordinate systems and a master clock; the universe is composed of infinitely many spacetime coordinate systems, or frames of reference. How fast the clocks tick depends on the relative motion between these frames of reference. Einstein's theory of special relativity was based on two postulates: all inertial reference frames observe the same laws of physics, and light travels at c.

The interesting implications about special relativity are not outlined in these two postulates, but rather, they arise from them. The concept of frames of reference is itself a marvel, because it suggests that we live in coordinate systems specified by spatial and temporal coordinates. Moreover, those coordinates may be transformed if our coordinate systems are moving at a significantly large velocity relative to some other coordinate system that is stationary. How large? Near the speed of light. How so transformed? By time dilation and length contraction, as per the Lorentz Transformations.

The reason why we do not see these effects (time dilation and length contraction) on a daily basis should now be obvious: good luck getting a banana launcher to move near the speed of light! It's no wonder our clocks don't actually get slower if we hop into a car and speed up, because the speeds we can attain are nowhere near what is mathematically sufficient to even be considered close to the speed of light. In the example above, I indicated that both the flying pigeon and the woman on the train move at 0.75c, but that's still 75% of 300 billion meters per second, which is a lot of meters per second![8]

What do you get if you aren't moving fast enough? We will still get relative motion, but the relative time we get will be so insignificant that it will look as though the difference in our clocks is essentially zero. Absolute time, and therefore Newtonian Relativity! If observers in frames of reference are moving slow compared to light, the effects they observe are reminiscent of what Galileo and Newton proposed, but when light is taken into account, this fails.

Special relativity is not a statement of contradiction to Newtonian relativity. It is a wider picture that takes into account extreme situations that we don't encounter in the everyday world. All Einstein did was state that light always moved at c and that all inertial reference frames must observe the same laws of physics. The latter part of his assertion was developed by Newton and Galileo, but this time, "laws of physics" didn't just mean mechanics. "Laws of physics" meant electromagnetic phenomena as well, because Einstein did not feel comfortable with making exceptions. According to Einstein, if mechanics obeyed Newtonian relativity, light must also

---

8   This is roughly 100 billion times faster than the speed at which New Jersey drivers go when they pull their cars out of parking spaces.

obey Newtonian relativity. If it did not, someone needed to write down a new theory of relativity, which is exactly what Einstein did.

When special relativity was first introduced, it did not receive the glamour that one might have expected it to garner. It introduced an idea that was so peculiar (time and length can both change, and they are intertwined!) that people dismissed it.

A large component of the issue was a dominating perspective of gravity. The standard notion of gravity of Einstein's time was Newtonian gravity. Newton proposed that gravity is a force that can be experienced between multiple objects. Every object with a measurable mass is subject to the gravitational force of some other object. The force between those two objects is directly proportional to the objects' masses and indirectly proportional to the square of the distance between them.

If an apple fell from a tree, it accelerates towards the Earth because Earth pulls on the apple with a gravitational force. The result is of course a downward acceleration of the apple that we can and have directly measured to be −9.8 meters per second squared.[9]

---

9    This acceleration would only be obtained in the absence of air resistance or any other forces that act in addition to gravity. The minus

The apple pulls on the Earth as well, but we would never see the Earth accelerate at +9.8 meters per second because the mass of the Earth is so large that, when compared to the mass of the apple, its movement is hardly seen.[10] We can make that kind of estimation and get away with it.

However, let's say we try and take two apples that experience a gravitational force between each other. Because each has a defined mass, they will pull on each other (regardless of whether we visibly detect that pull or not). If we try to alter the mass of one apple while keeping the force between them constant, then according to Newton, the distance between the apples must change. If you were to place the apples on opposite ends of the universe, there would still be a "wiggle" in the distance measurement. Maybe one apple moves forward a little, maybe it moves a little back. The essence is that the wiggle happens instantaneously, since the force value doesn't wait to be changed after a change in mass is introduced. Recall, however, that the two apples are at opposite ends of the universe. The act of changing the mass of one apple must have sent some sort of signal to the other one that tells it to start wiggling. That signal traveled the entire length

---

sign is included because the direction of acceleration matters in a frame of reference. The acceleration of gravity points downward, and we have established upward to be positive, meaning a minus sign for this acceleration is required.

10  The Earth is not displaced to any measurable extent but it still observes an acceleration in the opposite direction.

of the universe instantaneously, which means it has to transverse billions of light years in a time interval of zero seconds.

Therein is the problem. No signal can move faster than light, and Einstein was prepared to defend that until the end. But in order to do that, he needed to demonstrate that gravity was not what Newton thought it was. Einstein had no choice. If he desired a universally elegant theory, he had to refine special relativity to incorporate gravity as well.

# 6

# CLUELESS MONKEYS

---

Understanding the reasons for general relativity's relevance points to the recurring idea of the speed of light. Many physicists of the early 20th century wholeheartedly believed that no signal, object, or body could move faster than c. Proponents of special relativity (especially Einstein) argued that without this bold declaration, experiments done by Michelson and Morley could not be adequately explained. Moreover, without an adequate explanation of such experiments, previous mathematical work of physicists like Maxwell would not be empirically explained either, which meant yet more experiments done by people like Thomas Young could not possibly sit well with members of the physics community. It was very much a tense situation.

Einstein envisioned the universe in such a radical sense that it quite literally baffled astrophysicists, particle physicists, and just about all other physicists. This was not because of the theory's technical detail or complexity, but because for the strength of its departure from "common-sense thinking," the amount of experimental evidence that has arisen to support this idea is remarkable. A full appreciation of the elegance of general relativity has not been attained as of yet because the predictions it makes about what we might discover, coupled with the contradictions it introduces with what we already have discovered, are growing. It is a beast of modern physics with high-level mathematics thrown around as if it were conversational slang. Nevertheless, it has been regarded as the "most beautiful of theories." I agree; maybe you will too.

## "c"

Shortly prior to 1915, there was a question about gravity concerning light's speed. Essentially, Einstein reasoned that if light moves at the fastest possible velocity, which is precisely not instantaneous, then instantaneous communication between objects that experience a shared gravitational force is not empirically possible. This is because according to the gravitational force idea, adjusting the masses of and/ or distance between two objects can theoretically induce an instantaneous change in the force between them; if those objects are placed more than three light years away, for example, then the signal that triggers the force change must move faster than light. We have already established historical opposition to this possibility.

And therefore, the Newtonian view of gravity – the one that incorporates a gravitational field that induces particle interactions we call forces, is flawed. Yet, if you drop an apple from a tree and calculate its acceleration based on Newton's force laws, you will find that its measured acceleration is remarkably similar. Astronomers can and have predicted the orbital velocities of stars based on Newton's force equation. On a moderate level of observation, Newton's law has been supported by experiment. Only when extreme velocities are considered (by which I mean the speed of light) does Newton's law break down; this idea itself merited examination from Einstein's part. Dismissing Newton's view of gravity did not just mean eliminating the only credible explanation of gravitational phenomena, but it also meant dismissing a theory that was reasonably supported for over 200 years. You can therefore imagine that writing down general relativity most certainly required a fluency for theoretical considerations, but it also brought out incredible audacity and confidence from its author.

General relativity refers to the idea that space and time are intertwined into one coordinate system but it gives it the name of Minkowski spacetime. You take three spatial dimensions and add a temporal dimension, such that for any event in spacetime, there are four measurements you can make: where it is along one axis, where it is along another, where it is along yet another, and when the event happens in the first place. Minkowski space permeates all of what we consider the universe, even as it expands. You can think of it as the universe's spacetime map.

General relativity states that the spacetime map, if all the matter in the universe is taken out, is empty space. With an empty universe, all that is left is an infinitely large "sheet" of spacetime, and only when matter is introduced onto the sheet does it fail to remain flat.[11] If a banana were placed on the spacetime sheet, curvature would be induced but it would most certainly not be as strong as the induced curvature from a boulder, for example.

One of general relativity's key postulates is that spacetime curvature is directly responsible for producing gravitational effects on objects. Meaning, there is no such thing as a gravitational force but rather, the curvature of spacetime that causes planets to be pulled into orbit and apples to fall from

---

11 The description of spacetime as a sheet is not in accordance with its described dimensionality. According to general relativity, spacetime has four dimensions and sheets have only two. For purposes of envisioning these ideas clearly, one can think of empty spacetime as a sheet. The spacetime sheet is responsive to matter and it responds by curving. The degree to which spacetime curves is dictated by how much matter is put on any location in the sheet.

trees. If you learned that the gravitational force is defined by some inverse square law of distance, which is exactly what Newton says, Einstein says you learned it incorrectly.

As an example, consider the universe as a giant spacetime trampoline. If you were to place a heavy bowling ball in the center of the trampoline, it would form a valley on the trampoline's surface. If you were to then throw a cherry onto the trampoline, the cherry would travel into the valley and circulate inside because the trampoline is curved. An innocent observer on the side declares that the orbital dynamics of the cherry suggest that it is subject to the gravitational force of some other component in this trampoline-cherry-bowling ball system. A Newtonian approach would incorporate thinking along the lines of:

"Because gravity is produced by a field, the interaction between the cherry and the bowling ball is governed by forces. The reason why the cherry orbits the bowling ball is because the ball exerts a gravitational force on the cherry that provides it with the necessary acceleration to stay in orbit."

General relativity instead argues that the reason why gravitational effects are observed in this system is not because one component produces a force on another component. The bowling ball and cherry are matter that sit on a larger Minkowski spacetime plane which can curve in response to the very presence of these two objects. When the ball enters the plane, there is enough curvature to pull the cherry into its observed orbit. There is no need for a field, nor a force, nor any sort of inverse square of distance calculation. Einstein's view of gravitational dynamics is centered around the idea that gravity is not produced by the dynamics of a spacetime system. Gravity is the dynamics of a spacetime system.

Which side you take is dependent on what you think about light because that's really what this contention boils down to. Supporting special relativity does not require subsequent support for general relativity, but the latter was pragmatically invented to generalize gravity arguments concerning the former.

General relativity does not entirely eliminate all issues of light. For example, you might ask the question: "if spacetime is curved by matter, and light travels through the universe, are the trajectories of light also curved?" Certainly, light travels through spacetime but if spacetime itself is curved, you can no longer say light travels in straight lines. Calculations in optics allow us to predict where light will travel after being reflected from a surface based on the assumption that it will travel in a straight line. How does this contradiction fit in with general relativity? Moreover, general relativity states that spacetime curvature affects matter, which by definition has mass. This means if light can be affected by spacetime curvature, it must have mass.

## A CURIOUS EQUALITY

Newton's two masterpieces, the laws of motion and the law of gravity, both involve separate "types" of masses. In the law of motion, $F = ma$, the "m" for mass is inertial mass. Inertial qualities are synonymous to resistive qualities, which fits in with the definition: the purpose of an object's inertial mass is to resist acceleration of that object upon being acted on by a force. However, in the law of gravity, there is a different kind of mass known as the gravitational mass. In other words, how much mass an object has, in the gravitational context of Newton's law, is how much it pulls on another object. Before Einstein came along, it was accepted that these two masses are not equivalent. Such an equivalence argument did not seem relevant because gravity dealt with objects that occupied astronomical scales. Why should it matter?

It most certainly mattered to Einstein. He declared that the inertial mass of an object is no different from the gravitational mass of an object. It is appropriately referred to as the principle of equivalence.

Consider this: there are two monkeys in separate rooms with banana trees in each room. Monkey A's room will be placed in a gravitational field that produces a downwards gravitational force on all objects in the room. This force is responsible for a gravitational acceleration, which we can call g that all bananas, regardless of mass, will have if they were to fall from the tree. Recognize that if bananas fall downwards, we have to include the sign in our acceleration measurement. Taking downwards to be negative, the true acceleration of the bananas is -g.

Monkey B is in a room that is moving up into space with the same acceleration g, but it is not in a gravitational field. This means that if a banana were to be released from a stationary tree in a room that is moving up, the room's floor would rush up to meet the banana. It might look as though the banana were falling to the floor but it would obviously depend on the frame of reference. The room is nonetheless moving through space at g, which means the banana falls to the floor at -g. This means there is a tremendous physical difference between both rooms (one is in a gravitational field, one is simply accelerating through space), but a banana that falls from a tree in either room falls at the same acceleration.

Let's say in exactly ten seconds, bananas are going to fall from trees in both rooms. Both monkeys are going to watch their respective bananas fall and they end up measuring the acceleration to be -g, as expected. However, if you were to ask any given monkey how the banana fell, there are two possible answers it could give you: "my room isn't accelerating anywhere but there is a gravitational field" or "my room doesn't have a gravitational field but it sure is accelerating." We know that the latter response should be Monkey B's answer and that the former goes with Monkey A. In reality, they have no idea what to tell you.

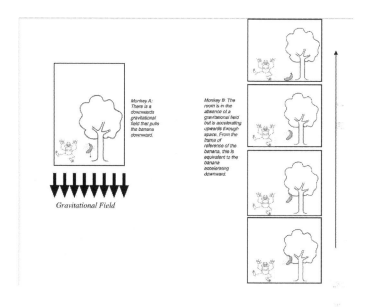

Monkey A: There is a downwards gravitational field that pulls the banana downward.

Monkey B: The room is in the absence of a gravitational field but is accelerating upwards through space. From the frame of reference of the banana, this is equivalent to the banana accelerating downward.

Gravitational Field

It is similar to Galileo's ship, in which you could not detect if the ship was moving because you were in an inertial reference frame. The implication is that you observe the same laws of physics as you would observe if you were in a stationary reference frame. Einstein therefore postulated that it is impossible to detect an experiment difference in either an inertial (no acceleration) reference frame with a gravitational field or a non-inertial (accelerating) reference frame with no gravity. The condition is that both reference frames must be accelerating at the same value, for example, -g. The only way this postulation has any sort of justification is if the mass of a body in the context of inertia is the same as the mass of the body in the context of gravitation because both are associated with accelerations that are relativistically indistinguishable. Thus, inertial mass must be equivalent to gravitational mass.

In order for these monkeys to truly not know what is going on, they must be only allowed to look at their immediate surroundings in spacetime. Meaning, they cannot be in gigantic rooms because that is too much volume for immediate localization. Consider, for example, taking a Monkey C and putting him in a room above the Earth. Monkey C's room is large enough to span to one eighth of the Earth. It has two trees on opposite ends of the room. If one banana is dropped from each tree, does Monkey C still not know what is going on? The bananas are going downward but will they necessarily fall in the same direction?

Take a pizza slice and hold it front of your face such that you are staring at the face of the crust. Take a strong magnet and put it at the tip of the pizza slice (facing your feet) and take two magnetic marbles that you can hover over opposite ends of the crust. What happens when you drop the marbles? They will travel across the pizza slice to the magnet. Do they travel down the pizza slice at the same angle?

Of course not; they have to go in at different angles. If you asked Monkey C this question, it could immediately tell you that the two marbles are not falling straight down and are being pulled in from some source (the magnet). Similarly, in Monkey C's room, both bananas will be attracted to Earth at its center of gravity, which is at the center of the Earth. If the room is sufficiently large, the bananas have to travel at different angles in order to go to the center. At that point, Monkey C can say "I am in a gravitational field, this room cannot be accelerating because if it was, the bananas would have fallen straight down in the same direction." Monkey C would have guessed right.

Monkey C, who understands he is in a room that is subject to a gravitational field.

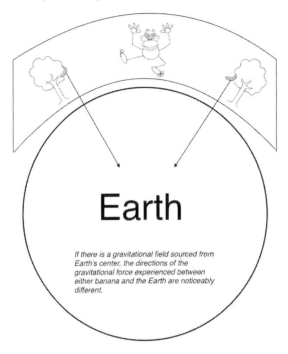

If there is a gravitational field sourced from Earth's center, the directions of the gravitational force experienced between either banana and the Earth are noticeably different.

A difference cannot be detected between gravity/accelerating reference frames in one's immediate surroundings. Why? Gravitational effects from spacetime curvature do not care about the masses of the objects they are acting on; they always give them the same acceleration, which is why everything in Monkey B's reference frame moves with him. However, if one looks past the immediate surroundings, the curvature of spacetime forces objects to take different trajectories to get to the same location, which is why Monkey C can tell the difference.

In other words, the acceleration from gravitational effects is indistinguishable from the acceleration of a force, as long as the two accelerations are the same (magnitude and direction-wise). Even if light does not have inertial mass, it can still be subject to spacetime curvature because in the context of general relativity, gravitational mass and inertial mass are physically equivalent.

It turns out, putting a blueberry muffin in space does distort spacetime but not enough to bend light. Large masses in spacetime curve the trajectory of light, which means light can be bent. Then how large does a mass have to be to do this? If you get a blueberry muffin the size of multiple suns, then you have yourself an object in spacetime that can bend light. At that point, the phenomenon you see is referred to as gravitational lensing.

Consider a star in a distant galaxy that is sitting at some point in spacetime, radiating light. If the path between this star and your eyes on Earth is not interrupted by any matter in the universe, light travels in a straight line, you'll be able to see the star as a result. However, when light does hit chunky matter that is large enough to produce a measurable effect, light does not travel in this fashion. As light approaches the chunky matter, it is in a straight line. However, the geometry of "chunky" forces it to travel along the chunky surface, which splits the light into millions of beams that take the shape of the chunky matter itself. If you have a baseball the size of one million cities put together in spacetime, light approaches the baseball's center and then is deflected by the curvature. It takes infinitely many paths along the edges of the baseball's surface, makes its way around the curved sections, and hits our eyes to produce the relevant two-dimensional shape. It is properly referred to as an Einstein ring. Varying the parameters in this theoretical experiment, let's say you want to make the baseball ten times bigger than that. At that point, what do you see?

You don't. Light can be bent but if the gravitational effects of spacetime curvature are that strong, then light itself will be pulled in. Sometimes, light can even be put into orbit around matter of this size. When light hits the surface of these large objects, the gravitational strength does not let it leave, meaning there is no reflection of light. These "massive objects" are called black holes (hence the light being pulled in) and they are explained beautifully with general relativity. Since light is the fastest moving object in the universe, if light can't make it out of a black hole, no other object can escape it.

## STREET CRED

Since 1915, there have been a number of cosmological anomalies that are perfectly explained by general relativity. The most primary is perhaps the Perihelion of Mercury's orbit. From Kepler's laws of elliptical orbits, we know that planets move around the sun in the shape of ellipses. In Mercury's case, however, at the point in its orbit that is closest to the Sun (the perihelion), there is a slight disturbance. Mercury, at its perihelion, does not follow an elliptical orbit. It is observed to move forward a little with each orbital round it takes.

Newton's law of gravity does not explain this observation, because there is indication in his equations of a disturbance in spatial orientation during orbit. General relativity accounts for this observation in an elegant and concise sense. At the perihelion, the Sun's distortion of spacetime is so strong that it forces Mercury to take a beaten-up path as opposed to a smooth elliptical one we would expect with Newtonian gravity. After all, Mercury is the closest planet to the Sun, which would explain why there is such a large disturbance in spacetime. It would most certainly be strong enough to alter Mercury's path. Urbain Le Verrier discovered this precession of the perihelion nearly one hundred years earlier than the publication of Einstein's explanation in 1919.

In 1916, Einstein submitted his work in general relativity to the Prussian Academy of the Sciences in the form of his field equations, which describe the distortion of spacetime by the energy and momentum of matter. It was widely regarded as the biggest triumph of 20th century physics and probably will be regarded as so in the 21st as well. To imagine the universe as a spacetime map with curvature properties and bent trajectories of light is impressive. To garner support from observations of gravitational lensing, black holes, and a perihelion precession is incredible.

As we will see, general relativity makes many appearances in a number of other physical phenomena and it has been assessed in over twelve experiments in the past hundred years. Thus far, general relativity has passed every test, including the ones currently being conducted. It is a potential, universal understanding of spacetime with unparalleled complexity.

# PART 3

## MODERN COMPLICATION

# 7

# THE MYSTERIOUS BANANA ROOM

———

Physicists can have it rough when confronted with mysterious problems, not because they have any more issues than they would have if the problems weren't mysterious, but because of popular culture's tendency to convolute an understanding of the mystery. It is no surprise that the media and movies can twist science when science does not specify precisely what it is talking about. It has been speculated that there is a different type of matter in our universe that accounts for certain mysterious observations. Galaxies have been heavier than they should be, stars move faster than they should move, and particles collide unexpectedly. A string of explanations has been provided by research scientists in the area of astronomy. In general, these explanations incorporate the existence of dark matter.

The reason why dark matter is "dark" is that it is invisible to the human eye and to every telescope technology built thus far. What that means for research purposes is that dark matter cannot be as easily experimented with, as can particles that fall under the definition of "matter." If you want to test Newton's laws of gravitation, you can take a banana, put it next to Venus, and measure the gravitational force on the banana as a function of whatever variable you want. You can look at the banana's orbital acceleration, for example, but you cannot find yourself a dark banana, put it next to Venus, and understand how it behaves in response to varying physical parameters.[12]

The universe consists of 27% dark matter, which poses a very interesting question: how is it that over one-fourth of the universe is comprised of something we can't even see? Moreover, how has there been so much experimental support for something we can't work with?

---

12  No, it does not matter how ripe the banana is.

## A DARK MATTER GALAXY?

The answer is that you don't need to see something to "do science" on it. Two branches of physics in particular have proven very promising areas of experimentation: astrophysics and particle physics. Very recently, a team of researchers led by Petier van Dokkum at the W. M. Keck Observatory in Hawaii discovered an intriguing galaxy named Dragonfly 44. Dragonfly 44 has a very small collection of stars. Van Dokkum's objective was very simple: he knew the stars would be moving at some orbital speed. Based on previous literature, however, he also understood that certain stars in Dragonfly 44 moved far too quickly for what their true speed should be, according to calculations. The strategy was to evaluate what the orbital speed was supposed to be and measure what the speeds of the stars actually were. Comparative analysis between these values might reveal interesting (or maybe not!) possibilities.

The expected orbital speed of the stars is not incredibly difficult to calculate. There are two theories of gravitation that can be used to do so: Newtonian's law of gravitation or Einstein's theory of general relativity. According to the former, the more matter there is in a system, the larger the magnitude of the force on each component. The larger the force, the greater the acceleration, and the greater the change in orbital speed. According to the latter, gravity is related to spacetime curvature but there is a similar argument: more matter in the spacetime map leads to more curvature, meaning stronger gravitational effects. Both situations postulate that the more matter there is in a given area of space, the greater the change in orbital speed should be of the components (stars) that make up that system. Because there were relatively few stars in Dragonfly 44, it was assumed that the stars could not be moving very fast. However, when they calculated experimental orbital speeds, they ended up with values that were significantly greater than what they should have been. It doesn't matter which team you're on, Newton or Einstein, because you're ultimately going to come to the same conclusion: if these stars were moving faster than they should be moving, there has to be more mass in the system. Precisely, mass that scientists before Van Dokkum could not detect.

It fit the dark matter description quite nicely. If Dragonfly 44 was composed of dark matter, it would have mass that we can't measure. That extra mass might contribute to gravitational effects experienced by the stars, which is why they move with such fast orbital velocities. The next objective was to dig a bit deeper: if it could be calculated, how much mass was necessary to force the stars to move at their speed, it could be known how much dark matter is present in this galaxy. It turned out, Dragonfly 44 is composed of dark matter by nearly 99%. Van Dokkum and his team found a dark matter galaxy.

There is a particle accelerator at CERN laboratory that collides large hadrons, or a specific class of particles. Despite nearly a decade of debate about a unique name for the collider, participating physicists settled on the Large Hadron Collider (LHC). In the LHC, particles can be crashed into each other to produce collisions. When two particles collide, there are two variables that are important: the energy and momentum changes of the collision. Generally, in ideally isolated systems, both energy and momentum of collisions are conserved. That means if you send two bananas crashing into each other and watch the two bananas stick together after colliding, the energy of one banana added to the energy of the other banana will give you the total energy of the stuck-together bananas at the end. The same applies for momentum.

Yet, that's not always the case. If baryons and fermions, members of other classes of particles, are sent moving towards each other, momentum and energy conservation is generally not observed. In fact, energy and momentum tend to both be lost. However, if you force the baryons and fermions to collide ideally, meaning you deprive the system of everything but these two particles, you should get momentum conservation. That's exactly what has happened at the LHC: baryons and fermions have been put in isolated systems with the momentum before and after the collision both measured.

What should have been no difference became a very obvious difference: momentum was lost.

It was, therefore, postulated that dark matter particles were produced in these collisions. Such dark matter particles carried away some momentum to produce the change that was measured. Because dark matter can slip from detection, necessary measurements to verify these postulations could not be made directly. When the momentum of these supposed dark matter particles was measured and compared to the momentum change of these particle collisions, the data were exceptionally telling.

## COSMOLOGICAL CONSTANCY

As far as where dark matter studies are today, most of the research is focused on designing better methods of detection. If cosmologists can design direct methods of experimentation, as opposed to indirect measurement through particle collisions, they might obtain a more panoramic understanding of what dark matter is and where it exists.

Now here's a lingering question: if 27% of the universe is composed of dark matter, what is the other 73%? I'll give you the interesting statistic first: about 5% is matter. Everything composed of mass we can directly experiment with only comprises 5% of the universe. The other 68% is something even more peculiar than dark matter.

When Einstein first published general relativity, he made an important note. When matter sits in spacetime and curves it, it does not curve it permanently. It can contract the spacetime map in different directions but that contraction should logically never become accelerated contraction. Otherwise, the universe would appear to shrink. Einstein reasoned that there had to be a restoring force to such accelerated contraction. There had to be something that pushed space out so that gravity couldn't pull space in.

Einstein called it the cosmological constant and it referred to the energy density of a region of spacetime. The idea was that all areas of space have the same amount of "restoring energy," which is responsible for restricting gravitational "contraction." The ultimate result is that space is designed to be static. This is the core of Einstein's cosmology concerns: spacetime does not expand.

You can therefore imagine the shock and surprise of Einstein when, in 1929, the Hubble telescope discovered that galaxies outside of the Milky Way were moving away from each other. Moreover, in 1998, two teams of physicists (Saul Perlmutter et. al., and Brian Schmidt and Adam Reiss) discovered from observations of supernovae that the universe is accelerating outward. Both of these discoveries provided support for a critical observation: the universe in fact is expanding. Einstein immediately removed the cosmological constant idea from general relativity's discussions and called it the "absolute worst mistake" of his career.

According to the experiment and the Hubble's findings, Einstein was wrong about what he initially thought to be obvious. And the interesting part about what Einstein thought to be obvious is that in a sense, it is fairly obvious. Nothing in the Newton's laws nor in Einstein's field equations provide a justification for the universe's expansion. It doesn't make sense for spacetime curvature to create more spacetime that can be curved. It also doesn't make sense for gravitational forces to propel the universe outward.

To stress this idea, imagine an empty room that is environmentally static. You can't look into the room, but you can throw an unpeeled banana in through a small window. When you do, the banana lands in the room at a random point. Two seconds later, it comes flying out through another window, but this time, the banana is peeled. You've managed to peel a banana by throwing it into an empty room with nothing in the room to peel the banana.

Logically, you recognize that this isn't really possible if the room is truly empty. Bananas don't just peel open whenever;

there is a certain amount of energy that goes into a pressure field, which provides the impetus for the banana to peel open. Without this energy, the banana has to stay closed. You reason that this room does indeed have such energy. However, when you send a team of monkeys to go find the energy, they find nothing. You ultimately have to conclude that there is an energy fluid in the room, but it's invisible, undetectable, and nonreactive.

Let's say you do throw in another banana and it comes out peeled. You don't know where the banana ended up inside the room. All you know is that wherever it landed, there was a surrounding invisible energy fluid that forced it open. And if you send in thousands of bananas, you get the same results and make the same argument: the energy fluid was strong enough to peel them. The question becomes: what does that mean in terms of the distribution of the energy fluid in the room? If any random banana gets peeled open, you have to agree that the energy fluid at any random point in the room is sufficiently strong enough to peel a banana open. In other words, no part in the room is deprived of the energy fluid. It is not only everywhere. It is strong enough everywhere. If you want to make the jump, you can even say that it is constant everywhere, meaning the density of the energy in the room is not a function of position. It's reminiscent of Einstein's cosmological constant.

Quite frankly, it doesn't matter what the other characteristics of the energy fluid are. At least we know that the energy density is constant. We were given a task to come up with an explanation for why bananas get peeled in this room and that we did. The next task was to specify why it doesn't matter where the banana lands in the room and that we also did. The technicalities haven't been worked out but the framework is there. If this sounds like a boring end of the story, that's because it is a boring end of the story.

Equipped with this analogy, what causes spacetime expansion? Certainly, like the mysterious banana room, there may very well be an invisible energy fluid that forces the universe's edges to push outwards via a pressure field generated by the energy fluid. Better yet, as more spacetime is added to the universe's edges, the tendency to expand stays the same because every region of spacetime gets the same amount of the energy fluid. The fact that the energy fluid is invisible means that it is a different kind of energy: it is dark energy. This is what is postulated to comprise the other 68% of the known universe.

Dark energy studies are not as conclusive as dark matter studies are. With dark matter, experiments have been conducted since 1884 that provide evidence for a type of matter that accounts for undetected mass. With dark energy, that's not necessarily the case. Research in the field has only emerged in the last thirty years and even then, it hasn't been as promising as people might think.

In fact, very recently, Professor Subir Sarkar at The University of Oxford has provided refutation for dark energy postulations. Professor Sarkar has argued that the data might fit the model but the data isn't broad enough to make significant conclusions. For one thing, the majority of research in dark energy has concerned supernovae, which are exploding stars that release energy. By making energy measurements of supernovae, physicists have been able to measure the distance the supernovae are from where they measure. The change in distance gives a potential understanding for how far the stars have expanded outward. According to Professor Sarkar, the problem is that the supernovae under experimentation have been localized to one area of the universe. Data from one area shouldn't be used to predict results for the entire universe.

Furthermore, the data that came from the supernovae under experimentation isn't very precise. Instead, Professor Sarkar argues that cosmic acceleration rates are actually unequal in various areas of the universe. The cause of those inequalities is due to a concept known as bulk flow, not dark energy. But you can pick whichever side you want to go with, because both are theories and not facts. That's the beauty of physics.

# 8

# HOT POTATO

---

In Dr. Ramamurti Shankar's Principles of Quantum Mechanics, he writes in the opening chapter why and how quantum mechanics contrasts notions of classical, comfortable reality. Relative to other mathematical arguments in the book, his description is straightforward: in observations of physical phenomena, there are "bubbles" that describe what observers of reality think about reality.[13] Quantum mechanics, special and general relativity, mass/energy equivalence, and ideas of the like are extreme cases of observation in which the classical "bubble" is too small to cover. Such extremes require bigger bubbles that surround the smaller, classical bubbles. If you ever want to find out how classical physics works, you can always use the bigger bubble of an extreme analog of

---

13   Dr. Shankar calls them domains, I call them bubbles.

the same observable. For example, relativistic motion for all speeds, is truly described by special relativity's arguments, but you can't go from a classical bubble to a quantum bubble, or from a gravitational force bubble to a spacetime curvature bubble, because those extreme bubbles require more substantial mathematical treatment.

We can apply concepts of mechanics, electromagnetism, and optics to engineering situations. We don't have to worry about quantum theory when we deal with tractors, because tractors aren't that small. At the macroscopic level, quantum effects are "washed out," analogous to how the interference pattern was washed out when a light detector is shined on the double slit experiment. You can't curve spacetime (noticeably) by suspending a baseball in the air, because baseballs just aren't big enough. Bananas and coffee cups obey the same laws of physics, but make the banana move at the speed of light, and suddenly, its clock slows down. Make the coffee cup gigantic, and watch a whole slew of photons bend as they hit its edges, forming an Einstein ring. Force the banana to become electron-sized, and suddenly, it's in the coffee cup, outside of the coffee cup, and everywhere in between, all at the same time. Until you look.

These questions bring us to a frustrating problem: why do we need to switch bubbles when we essentially are thinking about the same phenomena? Why should electrons respect superposition, but objects that are made of electrons fail to do so? Why is it that our universe can be thought of as a complex representation of very simple constituents, when the constituents themselves require entirely different laws of physics than do the objects which they constitute? Thus far, no physicist or philosopher has answered these questions. The fundamental frustration is that in order to provide an accurate answer, one has to develop a framework of physical laws that require no bubble alterations; meaning if you want to understand the way a classical system works, you don't need to simplify the quantum system down. You can simply apply the universal laws of physics to any system and get the same answer. Such an idea is popularly known as the Theory of Everything, in which every observable in physics must come along for the ride.

One of the issues is quantum mechanics. According to this beast of an idea, we shouldn't be able to simultaneously know the momentum and position of a certain particle at any given time. We also shouldn't expect to locate a particle, or to calculate a definite energy of a particle, unless we measure it directly. Why don't bananas act as waves? Why must energy be discretely "packed," while position measurements are allowed to be continuous? Because quantum mechanics specifies what happens on very small levels, and everything we know of is composed of many, many of these small levels, it should follow that everything we know of should obey the principles of quantum mechanics. Energy should be quantized, models should be probabilistic, and bananas should be able to collapse on a state of peeled, unpeeled, peeled half-way, or peeled by a fraction of 2/3.

But quantum decoherence is what prevents us from readily observing quantum mechanical effects on the macro-scale, which is why you don't need to define a wave function every time you want to know if a banana is peeled or not. Quantum decoherence occurs when molecules on the quantum scale lose their "quantumness" – this could be a result of radiation, bombardment by photons, or temperature changes. In the real world, light travels in all kinds of directions, and we are being hit by it constantly. Environmental radiation (and disturbances in general) aggravates the molecular perfection and placidity of the quantum world. The result is of course that the quantum world no longer behaves like the quantum world; it collapses on a classical state. Theoretically, this is why electrons produce the double slit interference pattern. Up until you measure where they are, they are everywhere. When you measure the system, you mess with the system, and the result is that the electron loses its quantumness. The system decoheres to an irreversibly classical state.

## THE FUNDAMENTALS

That classical world is a universe of forces and the motion of bodies that are subject to those forces. Meaning, if anybody wishes to apply quantum mechanics in a novel way such that it shares a direct, intuitive pathway with the classical world, classical forces have to be mathematically explained by quantum theory. There are known to be four fundamental forces in the universe, and every force acting on a random banana, on a Nascar driver in a race, and on airplanes during landing, is a combination of these fundamental four. They are the electromagnetic force, which describes interactions between charged particles, like the electron. The strong force describes interactions between very small particles known as quarks, which, when bonded "strongly," form protons and neutrons. The weak force describes the interaction between particles during radioactive processes, or when electrons, positrons, and neutrinos, which are very tiny, massless particles, split apart and are emitted from an atomic nucleus.

The fourth is the gravitational force. I'll guess what you're thinking: you read an entire chapter about how gravity might not be a force, so how is it now a fundamental force? When taken into the context of general relativity, gravity is not a force. Gravity is the curvature of spacetime, which produces gravitational effects that are analogous to forces, but are not. It is inherently different, to say the least, from the other three. However, for purposes of the fundamental force model, understand that you can treat spacetime curvature as the cause of force-like effects. Spacetime curvature is responsible, according to general relativity, but what it is responsible for is the effects we observe that remind us of a gravitational force. Namely, Newton's law of gravitation. Newton's masterpiece was held precedent for nearly 300 hundred years prior to general relativity's publication. Just because Einstein was the one to challenge doesn't mean the challenge should necessarily be taken as correct.

To describe how particles experience forces, we cannot simply state, "this particle right here has a force on that particle right there, do the math and you get the answer." Particles can experience forces in different ways, depending on the force. In Maxwell's day, the idea of a field emerged to explain the electromagnetic force. A field is essentially a set of imaginary lines in spacetime, which originate and fly through random directions in our universe. Fields are responsible for "acting" on a particle that is placed in them, and the result is a force from those fields. Maxwell reasoned that spacetime was permeated with an electromagnetic set of field lines, such that if a charged particle were to be placed in the field, it would experience a force. The motion of the particle could then be deduced from the force.

For example, if you take two charged bananas and put them next to each other, they will experience an electrostatic force between each other. Surrounding the bananas, in the room, or container, or whatever banana holder they might be resting in, is a sort of banana smoothie. This banana smoothie is continuous, and although composed of tiny granules of banana, it really is just a fluid medium. Electromagnetic field theory states that if you simply took one banana and put it inside the banana smoothie, there will be a value at whatever location you place the banana. There will be a force on that banana, which happens to be proportional to the value prescribed by the field, or in this case the potential, and the charge of the banana itself. This is not analogous to keeping bananas in an empty box and waiting to see what happens; there has to be a field, or in this case, a banana smoothie, that makes it happen.

Mathematically, you can also deduce the motion of charged particles with the Lagrangian operator, an operator that gives the action of a particle in the electromagnetic field. Using advanced mathematics to calculate the minimum action required to go from point A to point B, for example, you can then use an equation set known as the Euler-Lagrange equations to derive the profile of motion for that particle. Such methods are encapsulated in what is referred to as a classical field theory, and since we are talking about electromagnetism, the classical theory of the electromagnetic field.

If you're a fan of Newtonian gravity, in which gravitational effects are pure forces, you can say that a gravitational field exists just as an electromagnetic one does. The process is the same: put a thing with mass in the gravitational field, measure the interaction of "thing" with field, and find the motion.

These fields don't work for the other two forces, however. The strong forces and weak forces are instead described by Yang-Mills fields, which operate a bit differently. Though in essence, you can assume they get along nicely with classical field theory without any complications. If you're willing to dismiss general relativity, all fundamental forces are accounted for by classical field theories. However, there are two problems: the first is that we haven't talked about quantum mechanics yet, and the second is that we can't just dismiss general relativity once we've accepted special relativity. So if you want to dismiss special relativity too, then you can no longer use the relative motion idea of Newton either, meaning the entire framework you have established for mechanics is wrong. We do want to unify quantum mechanics with these four classical field theories to describe the motion of particles. Yet we also want to make sure that fits with Einstein's ideas as well.

The former issue has actually already been solved. Classical field theories define continuous fields in spacetime, meaning the lines in the universe are smooth and infinitely long. In quantum mechanics, however, all quantities are essentially quantized. So if you put an electron in an electromagnetic field and measure the energy, the value you get according to classical field theory can be whatever you want it to be. But quantum mechanics tells you that energies are quantized. The banana smoothie model of the electromagnetic field no longer works: you have to find a way to quantize a banana smoothie. In other words, field-particle interactions must obey characteristics of quantization. The same (well, a similar) case exists for the other three fundamental forces.

The need for something better than classical field theory becomes obvious. According to classical field theory, field lines are not localized, energies are allowed to be whatever they want, and you don't have a restriction of quantization. Quantum mechanics talks about the same particles that classical field theories talk about, but quantum tells a much different story.

## FRUSTRATIONS

To better understand this distinction, consider an example in which you and your friends play a game. In this game, you each are equipped with a measuring device that can take in a certain volume of banana smoothie and tell you how many bananas worth of smoothie you have. You take two bananas, mash them up, feed it to the device, and it reads, "2." You take a banana and a half, mash it up, it reads, "1.5." Take ten and one ninth of a banana, it reads, "10.1112." You get the point.

You all go into a room with infinitely many containers, and each container has a random amount of banana smoothie inside of it. You are allowed to toss such containers around at each other. Everybody will accumulate a great many number of containers, and eventually, your goal is to feed the banana smoothie in all of your containers into your device, and figure out how many bananas worth of smoothie you had in total. At the end of many rounds, the person with the highest number of bananas, as read by his/her device, wins.

You're probably thinking that this isn't too much of a game, it's really just a hassle. You have to deal with many containers of banana smoothie, pour all of it into a device, and keep count of how many bananas you have. The interaction between you and banana smoothie – or the job you have to do to get it into the device – is made possible by you (obviously) and by the banana smoothie itself. This is analogous to how classical field theory describes the interactions between particle and field: the field is continuous, and the behavior of particles in the field dictates the way in which they interact. The field is the smoothie, you are the particle. So ultimately, you can make measurements of how many bananas you have, but those measurements can be anything. I never said there was an integer number of bananas in each container. Realistically, if each container has a random amount of bananas, it could be anything! You can obtain 112.3453 bananas, 4 bananas, or 987.4774 bananas. It's a continuous domain.

However, let us say Werner Heisenberg walks into the room and demands that the number of bananas on all of your devices must come in integers, meaning your device can't read 2.5 bananas worth of smoothie, or 5.7 bananas worth of smoothie. The device can give 2, 3, 5, 6, or any other positive integer. But it has to give an integer. Also, Heisenberg did not say that the containers had to have integer number of bananas, meaning the containers will still be random. Your device is the one that has to give an integer value.

Now it probably takes a lot of self-restraint to not tell Heisenberg to get out of the room and never come back. After all, somebody made banana smoothie in an infinitely large container and poured out random amounts in infinitely many containers, all of which went in the room. How can you assume that each container was given an integer value of bananas? If you don't believe me, you try it; blend up one hundred bananas and try and get exactly one banana's worth of smoothie in one hundred separate glasses. It is highly unlikely. However, Heisenberg maintains his interesting requirement; however many bananas you measure at the end of this game must be quantized. Heisenberg demands quantum banana mechanics, and you have to make it work.

How do you do this? Can you still play the game with banana smoothie? Of course you cannot, there is no way to guarantee that everyone gets a whole number of bananas' worth of smoothie at the end of this game, especially if you have many, many friends. So what do you do?

You change the game such that you and your friends are no longer passing around banana smoothie containers, you are passing bananas. As long as you pass around whole bananas, Heisenberg's restriction fits. You can play for ten years, and you would all still obtain an integer value number. You have successfully created a quantized banana game.

What you just did is define a new parameter for the method of exchange in this room. You understood that banana smoothie cannot be used if there is a restriction of quantization, so you quantized what you passed around to make it all work. Similarly, classical field theory cannot obey the laws of quantum mechanics if the fields are continuous. Their interactions with the particles inside of them must be quantized, similar to how the banana smoothie containers must be replaced with bananas for the restriction of quantization to be maintained. In order to unify quantum with the fundamental forces, and subsequently the field theories, quantum field theories become necessary. Does this mean we need to get rid of banana smoothie joints everywhere in spacetime? No, it just means we have to define interaction in field theory as quantized interaction. No longer do continuous domains apply.

In the latter portions of the 20th century, quantum field theories for three of the four forces were defined. The first was a quantum field theory for the electromagnetic field, or Quantum Electrodynamics (QED). QED was pioneered by many physicists, but the most popular was Richard Feynman. He published a series of lectures in the 1970s that gave a non-mathematical explanation of the subject. Essentially, the electromagnetic field indeed agrees with quantum-mechanical observations, because electrons interact with each other through the emission and absorption of photons. Any time an electron behaves a certain way, it is because it has absorbed a photon or released a photon. The electromagnetic field is quantized, because photons themselves have quantized energies. There still existed an electromagnetic field in spacetime, but its continuity is irrelevant in terms of the interaction between particles that were located inside of it; the interaction, and subsequently the observed forces, is entirely due to the exchange of photons, according to QED. The banana smoothie nature of the electromagnetic field is replaced with the whole banana nature of the photon, which means the electromagnetic field officially agrees with quantum mechanics. This was the first unification of special relativity, the extensive study of light, with quantum mechanics, the extensive study of particle and wave behavior.

Quantum Chromodynamics describes the strong forces between quarks. Quarks are very small particles located in the nucleus of atoms, and they experience strong forces between each other. Depending on the properties of the quarks in a collection of them, the strong forces between bind them into protons and neutrons. Quarks do not just interact with each other through a strong force field; they transmit the information of force interaction, similar to the way electrons do with photons, through "messenger particles" known as gluons. Gluons can be exchanged between quarks, and the mathematical attributes of the exchange process dictates, at least in QCD, the observed forces on the quarks in question. In this sense, the strong force can also be defined by a quantum field theory, because if the force is itself quantized, there should be no reason to preserve requirements of continuity in field ideas. The gluon in QCD is the banana, and it quantizes the fermionic field of the strong force.

Quantum Flavordynamics is a colloquial term for "electromagnetic-weak" theory, which defines the unification of the electromagnetic and weak interactions. Developed by Muhammad Abdus Salam, Sheldon Glashow, and Steven Weinberg, who shared the 1979 Nobel Prize for their work, QFD proposed that in high-energy physics, electromagnetic forces and weak forces converge into a single electromagnetic-weak interaction. Furthermore, the weak interactions could be described with their own quantum field theory, or quantum flavordynamics. Weak interactions occur between electrons, positrons, and neutrinos in radioactive decay processes, but again, these particles do not just interact freely. They exchange information for force interaction with particles known as W and Z bosons, which were also discovered by Abdus Salam et al. A parallel to QCD and QED, QFD defines a quantum field theory for the weak force. It does so by defining a banana: the W and Z bosons.

QED, QCD, and QFD all propose that the particles they describe – electrons, quarks, neutrinos, and positrons, interact with "messenger particles." There is no more continuous, infinitely long field with which you use your Lagrangian operator to calculate the motion of particles inside of it. Instead, particles interact by exchanging their own bananas, which, outside of the analogy, are known as gauge bosons. For QED, the gauge boson is the photon. For QCD, it is the gluon, and for QFD, it is the W and Z gauge boson.

Along with the conception of these quantum field theories came the development of the Standard Model of Particle Physics. The Standard Model is the first widely encompassing theory that describes how physics operates for multiple domains of particle behavior. It has been in development for nearly half of a century.

## PARTICLES, PARTICLES, AND MORE PARTICLES

There are two classes of particles in the universe: those that obey Fermi-Dirac statistics, and those that obey Bose-Einstein statistics. Particles described by Fermi-Dirac statistics, as outlined by Enrico Fermi and Paul Dirac, are appropriately referred to as fermions. They are the fundamental constituents of all matter, because they obey the Pauli Exclusion Principle. The Pauli Exclusion Principle states that no two particles can be defined by the same quantum numbers in spacetime. Quantum numbers can be colloquially thought of as a set of coordinates for quantum particles. The technical definition is the set of numbers defined to be solutions to the Schrodinger wave equation, which describes the dynamics between the wave function of an electron and its energy. For electrons in atomic orbit, the quantum numbers are energy level, angular momentum, magnetic spin, and the direction of the angular momentum vector. Essentially, the quantum numbers of an electron tell you where the electron is likely to be located in space (this is quantum mechanics, it's probabilistic!).

What does it mean to have different quantum numbers? Let us say some particles are defined in some space, and you assign every particle a set of quantum numbers in that space. The Pauli Exclusion Principle tells you that if you find that two particles have the same set of quantum numbers, you are wrong. In other words, no two particles can occupy the same place at any one time. So if you ever have two particles that just can't be at the same place at the same time, those two particles fall under the category of fermions. That's where Fermi-Dirac statistics apply.

Electrons, protons, and neutrons are fermions. Muons (discovered at The California Institute of Technology, 1936), particles with the same charge as electrons but with 200 times the mass, are fermions. The tau (discovered at Stanford, 1971), is also a fermion; it too has the same charge as an electron, but with approximately 3400 times the mass.

Separately defined for the electron, tau, and muon are the electron neutrino, tau neutrino, and muon neutrinos. Neutrinos, in general, are particles with no charge and relatively no mass (they have some mass, but very little). Neutrinos were hypothesized to exist as early as the 1920s, because in weak interactions, particularly with beta radioactive decay, energy and momentum were discovered to be non-conserved. It was postulated that some particle exists in radioactive processes that carries some of the energy and momentum away with it. In 1959, Clyde Cowan discovered such a particle: the neutrino. There are three known neutrinos today, and each combines with its respective fermion to form leptons, a class of fermions that can exist freely, at any location in spacetime.

Another class of fermions is quarks. Quarks are small particles that interact with each other via strong forces to form protons and neutrons. In 1964, physicists at the Stanford Linear Accelerator Center discovered particles that were supposedly smaller than protons and neutrons. Up until that point, everyone knew that "atomic" was not as small as you could get, there existed a subatomic level that was comprised of protons and neutrons. Quarks are the opposite of leptons, which can exist independently; quarks must bind with each other via strong interactions to form larger particles. At the time of the quark's discovery, not too many people suspected there was a sub-subatomic level, which is precisely what the quark is. It was a revolutionary finding in particle physics.

Satyendra Bose and Albert Einstein, on the contrary, developed Bose-Einstein statistics, in 1924. Bose-Einstein statistics apply to particles that disobey the Pauli Exclusion Principle. In other words, if you take two particles that are known to obey Bose-Einstein statistics, you can theoretically have both exist at the same location, at the same time. One might pass through another, or vice versa. The only known particles described as such are bosons.

Bosons are particles that cannot interact with each other. Their sole purpose is to be absorbed, emitted, or otherwise transmitted between fermions. The interactions that result between these fermions are responsible for producing the observable effects of forces. There are two types of bosons. One is the gauge boson, which, as described above, can be thought of as a messenger particle of a quantum field theory. Gauge bosons can be photons, gluons, or W and Z bosons, depending on the quantum field theory in question.

Think of bosons and fermions in the context of the hot potato game. You again take a bunch of your friends and buy very hot potatoes. As you pass these potatoes back and forth, think about what happens in your minds. You might catch one and burn your finger, and then yell at whoever threw you it. You might dream up all the possible reasons for why you should never speak to that person again, you might throw the potato back, or you might make french fries. Depending on the people in the game, it is safe to assume that there will be a very diverse collection of emotional and physical reactions as the potatoes are passed around. Relate yourselves to fermions, the room you are playing in to the field, the potatoes to the gauge bosons, and the reactions you all experience to forces. Plus if you still count the number of potatoes in the game at any time, you'll get a whole number. In this sense, quantum field theories define how particulate gauge bosons produce effects on particles that exchange them. Everything is quantized and we can still measure the forces, which means the quantum field theory is successful.

The Standard Model thus successfully explains QED, QCD, and QFD as quantum field theories, thereby providing mathematical unification with quantum mechanics. It classifies particles into two sets, defines how they interact, and mathematically normalizes three of four fundamental forces. Unfortunately, there's still one of those four forces left.

# 9

# CURIOSITY, CONFLICTS, AND CHEERIOS

———

One goal of the Standard Model is to provide a framework for all particles in the universe, while explaining how quantum mechanics agrees with the forces that physicists can measure between those particles. Three of the fundamental four forces – the electromagnetic, strong, and weak – "agree" with quantum mechanical uncertainty relations. Recall what these uncertainty relations postulate: that given an electron in an arbitrary region in spacetime, one cannot possibly calculate the momentum of the electron while simultaneously understanding its precise location in the spacetime region. The other is one I never introduced in previous chapters; it is a parallel uncertainty relation but it instead involves energy and time. If one were to force two electrons to exchange a certain amount of energy, the larger the amount of energy exchanged, the smaller the time interval of the exchange process. This means very high-energy particle interactions occur with increasing rapidity until energies ultimately approach infinity. If one were to lower the energy of the interaction to make measurement of the energy exchange experimentally feasible, one would sacrifice the ability to observe the process in an experimentally feasible amount of time (it would likely take too long)!

What this means for force interactions is embedded in special relativity's description of the cosmos as a four-dimensional spacetime map. If a cubical region in spacetime were considered, the region would have a certain amount of spatial and temporal volume. Placing an electron in the region means two observables can be measured: the electron's energy and its momentum. Decreasing the size of the region means decreasing its spatial volume, which increases the likelihood of you correctly guessing where the electron is. However, from the uncertainty relations, this means the electron will "buzz" ever the more frantically as its momentum value becomes more uncertain. Similarly, a smaller spacetime region necessarily implies a smaller temporal volume, which means high energy interactions are expected between particles that are in the region. If one were to consider any arbitrary spacetime "point," the energy and momentum observables would form a flying circus of rattling craziness.

You can therefore imagine the magnitude of confusion physicists would feel, if a different theory proposes that spacetime on a fundamental level is smooth and infinitely continuous; that said theory has been supported by observations of black holes, orbital anomalies, and gravitational lensing; that the essential purpose of the theory was to develop the veracity of special relativity, which, on account of the work of Paul Dirac and Richard Feynman, was integrated into discussions of quantum field theories in general. The very extension of the core arguments in quantum field theory attack the core arguments in a violent loop. This contention is due in full to two frameworks that do not get along: general relativity and quantum mechanics.

This apparent contradiction inherently accepts general relativity as the "more correct" perspective of gravitational effects in our universe; ever since Paul Dirac wrote the Dirac equation, or when Feynman and his colleagues developed quantum electrodynamics, the relevance of a spacetime multiverse in which Lorentz contractions act has become critical. Direct application of Newton's idea of a gravitational force is therefore no longer compatible. Unfortunately, the apparent solutions to be arrived at are no simpler than the problems themselves: either Einstein is right about a smooth spacetime and uncertainty principles require closer examination, or the latter stands, leaving a new theory of gravity to be developed.

If gravity is some sort of universal trampoline, how do you explain a quantum field for it? Quantum fields should span spacetime, so if there is a gravitational quantum field, does it span spacetime? This question contradicts itself, because the gravitational field cannot produce gravitational interactions and span spacetime, because the gravitational interactions are defined as the dynamics of spacetime. This is kind of like saying, "I have a collection of bananas right here that I will now drop on this entire table, but the table is also made of bananas." There might be a mysterious understanding of forces that induce spacetime curvature, but that would mean we would have to find it before incorporating it into a theory of quantizing the gravitational field. So now we've come to a very prevalent area of theoretical physics research that has left physicists stumped for as long as Einstein's reign: a way to introduce a quantum theory for the gravitational field. Nobody has found such a concrete definition, which means there has not been a "quantum gravity."

The quest for a quantum gravity is a weighty one because there are many theoretical conditions that must be sufficiently satisfied before an idea is considered a valid candidate. Quantum mechanics is the reason behind the majority of these conditions, because in each of the other quantum field theories, there is a particle that "pushes" the corresponding force along. In general relativity, force interactions are very much passive; you need to send a photon from an electron to another electron to measure some sort of electromagnetic force, but general relativity postulates that the mere act of placing matter in the spacetime sheet produces a measurable gravitational force on some other matter in the vicinity. In the 1930s, few physicists understood this problem as well as a Soviet by the name of Matvei Bronstein, who hypothesized the existence of a "graviton." Analogous to the other gauge bosons, Bronstein declared that the graviton was responsible for propagating the gravitational force between objects. His key constraint, however, was that the idea of the graviton only applied when spacetime curvature was kept extremely minimal, such that the universe could be approximated to be flat. In situations in which the curvature was abnormally large (such as when you have a black hole), Bronstein argued that gravitons could no longer explain what was happening.

## ASTRONOMICAL CONSIDERATIONS

Recall that black holes are defined as matter in spacetime that has collapsed on itself. The result is tremendous spacetime curvature, which produces immensely strong gravitational effects that suck in all matter in the surroundings. The black hole is analogous to a large bowling ball the size of Manhattan, let's say, on the spacetime trampoline the size of ten Manhattans. You know that if you throw a banana onto this trampoline, it will not remain stationary. There is a bowling ball the size of Manhattan stuck in the center, which means there is curvature in the trampoline! The banana is going to fall through.

In fact, if the bowling ball were large enough, everything thrown into the trampoline would fall into this bowling ball valley because naturally, the curvature of the trampoline would be enormously emphasized. The anatomy of a black hole is such that not every region inside of it is characteristic of the "nothing escape" idea; the black hole pulls things in towards what's called the singularity, which is the single point responsible for the spacetime curvature we see. As we progress from the outskirts of the black hole to the singularity, we cross a region called the event horizon. The event horizon marks the line from which nothing can cross back once it has initially crossed forward, meaning once you step over the event horizon, you cannot return. For the sake of an example, let us say you crawl into the trampoline and pass the event horizon. The trampoline becomes steeper as you go through, so steep that it eventually leads to the point of singularity.

What happens at the singularity? There are two mathematical anomalies, the first being infinite spacetime curvature. At the singularity exists infinitely strong gravitational effects because everything is pulled to it, but since general relativity predicts gravitational effects are spacetime curvature, spacetime must also be infinitely curved at the singularity. What that means is unknown.

Additionally, it would appear as though black holes are massive phenomena that require treatment with Einstein's field equations. This is true. At the core of the black hole is a particulate singularity, which is postulated to be much smaller than the size of an object appropriate for general relativity's consideration. In fact, the fact that the singularity can be thought of as a particle necessarily implies that quantum mechanics falls into play, which means the structure of a black hole is segmented into two reasonable perspectives. Not only do these perspectives cause interest, but they also call for frameworks that have already been shown to be mathematically antagonistic: how can general relativity account for a black hole while quantum mechanics accounts for the singularity?

Even if there were to be a singularity that obeyed the rules of quantum mechanics, quantum mechanics still does not explain how to "pack" matter into such a small volume. Electrons are point like particles, but this does not mean we can give the electron an information set to hold onto as it moves through all of the possible states in its superposition.

You might ask why it is important for astronomical physicists to listen to quantum physicists when discussing black holes.

After all, quantum operates on the smallest scales and black holes are in Einstein's realm of the big, the massive, and the spacetime curvy. There should be no reason as to why tiny particles should agree with observations made on the cosmological scale. But black holes do provide a reason; while they are massive, there is one point of a black hole that should theoretically obey quantum mechanics. You throw a couch into a black hole, and it goes to the singularity, but throw a couch at an electron, and it's not going to keep the couch.

An entirely separate problem is the nature of spacetime itself. The spacetime trampoline is big enough to hold anything thrown onto it. It spans the entire universe, and possible multiverses, because there is no mathematical disagreement with it doing so. As long as it is maintained the relationship between the trampoline's curvature to the gravitational effects it induces on other particles, spacetime can be infinite. However, quantum mechanics requires spacetime to be quantized, or at least, some property of spacetime to be quantized. After all, the energy of particles and the forces described by quantum field theories are quantized and they span all spacetime. Since gravity is defined to be a dynamic property of spacetime, according to general relativity, then spacetime must be quantized. If you leave spacetime to be continuous and infinite, it doesn't agree with the quantum mechanical behavior of other quantum field theories. In order to have a quantum field theory for gravity, you need to quantize gravity, which of course means quantizing spacetime.

Though if the universe is expanding, that means at some point in the history of the universe, there must have been some initial, minimum amount of spacetime "units." The only way for the universe to expand is for the granules of spacetime to have multiplied. Here arises another problem: how do you start with a finite number of granules, as predicted by quantum mechanics, and get to infinity, as predicted by general relativity? If you don't believe me, buy one banana and continue to cut it into pieces until you have enough pieces to feed infinitely many monkeys.

It is a reasonable assumption that spacetime is an infinitely voluminous banana smoothie that gets larger at every moment. It is also reasonable to say that at some fundamental level, you can break that banana smoothie down into banana smoothie particles. Yesterday, there were fewer banana smoothie particles than there were today, which means if you take the difference in the number of particles from yesterday and today and divide that by the time difference between yesterday and today, you get the rate of expansion of banana smoothie particles. You can do this all the way until the beginning of the universe, but at some point, you have to agree that there was a minimum number of banana smoothie particles, because you find the rate of expansion of banana smoothie particles is consistently positive as you move through spacetime. Where did those initial banana smoothie particles come from? And how is that ever supposed to get to an infinitely large banana smoothie? You can't start with a few and get to infinity. The only way to do that is to wait for an infinite amount of time to pass.

We cannot put quantum mechanics and general relativity together if the universe is infinite according to one and finite according to the other. There needs to be compromise. A quantum gravity has not been postulated. In order to do so, the gravitational force must be quantized, as it is with electrons and photons or quarks and gluons. In order to do that, we need to (1) define a force to quantize, (2) define how the force will be quantized, and (3) define which particles experience the quantized force.

With the three quantum field theories that behave nicely, the simple emission and absorption of gauge bosons permits all sorts of interactions to occur between participating particles. But we know the graviton doesn't behave like that, so we postulate that the graviton induces spacetime curvature. Meaning every time the graviton is exchanged between matter, a certain set of information must be transmitted that quantizes the method of interaction, as the photons do for QED, as the gluons do for QCD, and as the W and Z bosons do for QFD. Unfortunately, an information-transmitting particle such as the graviton defeats the purpose of general relativity entirely. The beauty of the "most beautiful of theories" is that it doesn't require a messenger particle.

General relativity was proposed to account for the fact that gravitational effects in a system cannot be exchanged faster than the speed of light. This is why there is spacetime curvature. In defining gravity as a dynamic of spacetime as opposed to dynamics in spacetime, the problem Einstein sought to solve is solved. In Einstein's pursuit of solving this problem, however, is arguably the most mind-boggling idea in current physics research: the fact that if one attempts to formulate a theory of quantum gravity, quantum mechanics takes your cake and eats it, too.

Depending on how it is received, criticism can be a method to lock away the potential for an elegant idea to communicate its elegance. Criticism can also be a critical component of scientific progress. Arguably, how one views criticism dictates the effect it has on one's work; sometimes, the effect is profound, sometimes dangerous.

## LOOPS ON LOOPS ON LOOPS

Most of the time, criticism in physics is objective. It is a known argument that a theory can possess all of the mathematical elegance the theorist dreamed for it to possess. It can postulate beautiful arguments that bring revolutionary things to the table, redefine the shape of the table, and perhaps bring new tables themselves. Yet, as impressive as a theory may be, pragmatism always triumphs. The only factor that merits support for a theory is empirical evidence in the theory's favor.

Contrary to popular opinion, physicists do not speculate any whim they wish about the universe and expect people to agree. There is a reason why Einstein, Maxwell, Feynman, for example, have the name recognition and academic reputation that they do. At the time of publication, their work was considered a bit of a stretch, but as experiments were conducted and syntheses between observations made, their theories garnered support in a way that other theories did not. If one believes there were only three physicists who worked to craft an understanding of light's behavior and quantum theory, one would be hopelessly wrong. For every theory that caught wind under an observation made to support it, thousands of theories lost credibility. The idea of criticism, then, boils down to straightforward arguments: data is required to propound a theory to further steps. Nevertheless, this most certainly does not mean that a little bit of an imagination is a lot out of line. If you try naming an influential mathematician or physicist who has not posited something "crazy," you'd most likely have difficulty.

There has been a ruthless amount of criticism for the theories of quantum gravity because proposing and justifying data for their arguments has proven incredibly difficult. The fundamental problem is the composition of spacetime: is it smooth and continuous, or is it particulate after all? The technology that currently exists is not powerful enough to provide an accurate answer. However, the theories have stood for nearly thirty years (or more). They are not as empirically supported as special relativity or wave-particle duality. However, they are incredibly audacious and fun to think about.

Loop quantum gravity and string theory, although they resolve to address the same problem, are different in not only approach but in outcome. Loop quantum gravity is perhaps less known simply because it is younger and has not been published, cited, and presented as much as its counterpart, string theory.

Loop quantum gravity addresses the issue of quantum gravity by targeting the conflict between a gravitational field and spacetime. Before the 1980s, a physicist named John Wheeler speculated that at the fundamental level, spacetime was composed of a foam-like substance. Wheeler reasoned that spacetime could not simply be a four-dimensional plane that sits there and responds to matter by curving. His idea of spacetime was a geometric network of different foam-like elements; call them cheerios, for simplicity. He worked in tandem with another physicist named Bryce DeWitt to write down a wave function but this wave function did not correspond to an elementary particle such as the electron. This wave function was for spacetime geometry.

Precisely like the electron in the shopping mall, which collapses on a location once you observe it, Wheeler and DeWitt introduced the idea that perhaps the geometry of spacetime was in a superimposed state of many different possible geometries. Perhaps one geometric orientation of spacetime was for cheerio 1 to be linked to cheerio 2 and another could be for cheerio 3 to be linked to cheerio 1. Perhaps cheerio 2 may be linked to cheerio 3 and maybe cheerio 4 is linked to cheerio 1. The notion was spacetime is composed of infinitely many of these links of cheerios and the geometric orientation of these cheerios is not a deterministic property of the universe. Furthermore, the orientation of the cheerios' intersection are the sites of quantized "bits" of spacetime, and spacetime's many possible geometries are simply superimposed.

The wave function is called the Wheeler-DeWitt equation and it is regarded as the quintessential equation of loop quantum gravity. Although the solutions of the Wheeler-DeWitt equation were solved in the early 1990s, problems still remained. While there was a defined perception of the dynamics of spacetime, there was not much to be said about the dynamics in spacetime, such as curvature. In other words, how do interactions take place in spacetime? From other quantum field theories, something needs to be quantized and subsequently "traded" across a gravitational field for a quantum theory of gravity to agree with the Standard Model.

In a quantum field such as the quantum electromagnetic field in quantum electrodynamics, there exists the possibility that a random photon strikes an electron. At that moment, energy is transferred to the electron. This creates a disturbance in the nearby vicinity because the electron may vibrate, move, or oscillate. The photon induces, and the electron performs, a quantum excitation of a quantum field – when particles in the field are affected by messenger particles and subsequently create disturbances in their surroundings. Loop quantum gravity postulates that the gravitational field is quantized because a quantum excitation of the gravitational field is a loop. These loops are simply objects that are extremely small and shaped with loop-like geometries. At the fundamental level, if the quantum field of spacetime is excited, a loop is found floating around. On the classical scale of physics, where energies are low, things move slow, and black holes don't exist, these loop excitations can be approximated to be the particles that have been hypothesized to account for transmitting the gravitational force: if one starts at a loop and "zooms out" to normal observational scales, one arrives at the graviton, the postulated messenger particle of the quantum gravitational field.

The loops compose spacetime, so if spacetime were to be excited, the loops would be excited. However, loop quantum gravity defines the loop to be the quantum excitation of spacetime, which means loops are also the ones passing the message for excitation. If loops excite and loops are excited, loops must excite loops.

## POTENTIAL ANSWERS

This is kind of like saying, "A society of monkeys communicates with each other by forming circles of monkeys, because monkeys serve the two functions of receiving messages and transmitting messages." However, because the monkeys permit both processes that are critical to producing observable forces, they are the only components in the system that are fundamental in an observational sense. Where the monkeys are is irrelevant because the observation we make is that messages are received and transmitted, and the understanding we have developed is that monkeys are responsible for our observation. A similar argument is proposed by loop quantum gravity; if loops are the fundamental units of spacetime that permit the interpretation and transmission of "messages," there is no need for further specification of where the loops are or on which background material they sit. According to loop quantum gravity, there is theoretically no need for the idea of spacetime, or any surrounding field, for that matter.

There is simply an infinitely large connection of loops that excite each other and produce measurable gravitational effects. If we keep spacetime in the picture, then we have to answer a series of questions concerning what role it plays, where the loops are, and how the process of loop excitation works. The "loops excite loops" postulation gives us the ability to disregard "spacetime" entirely and instead assert that loops connected with other loops are the reason why matter observes the gravitational effects it does.

In the early 1990s, Lee Smolin, Ted Jacobson, and Carlo Rovelli discovered that spacetime can be thought of as what's called the Ashtekar connection field.[14] They defined this connection field as a network of intersecting loops that were strung together to form the tapestry of what we thought was spacetime all along.

If you are weaving a sweater and you notice the threads aren't weaved together tight enough, you can obtain some more precise equipment and string the fabric more uniformly. You can keep getting more precise, until eventually, you get what looks like a well-woven sweater that doesn't seem to be composed of threads at all; it looks smooth. However, if you have small enough fingers, you can take it apart and see that it really is composed of threads, and there are points where

---

14  Named after Abhay Ashtekar, a pioneer of loop quantum gravity.

those threads might even intersect. What you thought was smooth was really just a very precise network of particulate objects.

If you start with a banana smoothie that appears continuous, you can obtain an ultra-fine banana smoothie parsing tool that lets you probe the liquid. If you continue to probe on finer levels, you find that the smoothie is composed of a foamy material, which is composed of dense connections of banana granules, which are themselves particles. Similarly, if observations and theories have suggested that spacetime is continuous, it may be that current technology is not sufficiently precise to obtain a close enough look. Perhaps there are tiny, woven entities that make spacetime look smooth when it really is not. Upon "zooming out" of the loop network, one arrives at regions of spacetime that look ever the more convincing of a continuous manifold.

When each loop crosses another, a region of intersection forms. When networks of crossed loops cross other networks of crossed loops, what is formed is referred to as a spin network. Every possible spin network in the universe, when strung together, forms what's called the spin foam, and the spin foam is what proponents of loop quantum gravity call the fundamental material of spacetime. Gathering support for the existence of these loops has proven to be, arguably, the most difficult task of successfully communicating loop quantum gravity to other members of the physics community. Theoretical calculations have shown, however, that the geometries of the loops agree precisely with calculations made previously by Einstein concerning general relativity.

The network of intersecting loops forms the stage for transmitting and receiving gravitational effects, which means the gravitational force according to this theory obeys quantum mechanical observations. If the loop network can be quantized, which means defining intersections that represent "spacetime granules," the energy of the universe is successfully quantized and the principle of a particulate, vibrating universe is addressed. This means that for classical physics, we can assume the loop takes the form of a graviton and acts analogously in the gravitational field as photons do for electrons, gluons for quarks, and W and Z bosons for positrons, neutrinos, and electrons. Interactions between loops are mediated by the loops themselves, which means "loop curvature" becomes possible if the loops excite each other correctly.

In regard to approaching the quantum gravity problem, loop quantum gravity is passive in approach. The conflict exists between multiple frameworks including quantum electrodynamics, chromodynamics, and flavordynamics, but at close inspection, loop quantum gravity only tackles the gravitational force. In doing so, it is hypothesized by the theory's equations that integration of the fundamental forces into quantum mechanics is possible, plausible, and accomplished. What a marvelous result; a theory that approaches one dimension of a four-tiered situation and promises a solution for all instances of all tiers.

The mathematics of loop quantum gravity are built on Einstein's field equations and attempt to explain general relativity in the context of quantum mechanics. The unfortunate aspect is that unlike general relativity, loop quantum gravity is not supported by massive bodies of astronomical data concerning gravitational lensing and black holes. Unlike quantum mechanics, loop quantum gravity has not been justified in experiments such as the electron version of the double slit experiment. And unlike special relativity, loop quantum gravity does not have the cornerstone support of experiments concerning light's speed. Loop quantum gravity is elegant, concise, and direct in approach. "Street cred" is what restricts it from being accepted as the theory of quantum gravity. Physicists are not ready to accept loop quantum gravity as a "theory of everything," and while the theory has its proponents, many argue that this is for good reason.

## SO, WHAT NOW?

For now, unification of physical laws has not been empirically reached. However, strung through spacetime, particle interactions, and forces are ideas that are being harbored by the curious. As humans, we might be infinitely far away from learning the common thread of the universe's tapestry, or we might be closer than humans ever will be to understanding the machinery of the cosmos. What is certain is the passive function of interpretation. The idea of pioneering a comprehensive framework that explains the small, fast, big, and complex, is a theme in physics based on audacity and caressed by mathematical elegance. Ancient revisitations are complemented by novel research queries, old questions are pursued by new solutions, and the apple that fell on Newton's head might just find its way into different fruitful contexts. Through calculation, reflection, speculation, and writing, we slowly evaluate the score of an intense game enriched with history and lined with the silver of youthful vigor and wit. We might dream of thousands of ways to experiment with nature, and the best part about it is the rulebook is written by a different hand.

# ACKNOWLEDGEMENTS

———

I would like to offer my wholehearted thanks to my teachers and mentors at Watchung Hills Regional High School. I am grateful for the engaging instruction in science and mathematics from Mr. Salvatore Fazzino, Mr. Christopher Gibson, Mr. Michael Gangluff, Mrs. Michelle Nunez, Mr. Daniel Twisler, Mrs. Christina DiBartolo, Ms. Chelsea Seidenberg, and Dr. Suzanne Fenstemaker. Their profound encouragement and advice has been invaluable in my education.

I also acknowledge the empowering examples of my humanities teachers who showed me the importance of holistic, written expression. Mr. Ryan Murray, Mrs. Jessica Kelly, Mrs. Jana Battiloro, Mr. Ira Horowitz, Mrs. Jacquelyn Gilliam, Ms. Kristin Czajka, and Mrs. Rebecca Brown have provided me with valuable insights on pursuing knowledge through writing and language.

My special thanks are extended to the kind, tirelessly focused people at New Degree Press: to Anastasia Armendariz, for her continued support in developing the manuscript, to Leila Summers, for her persistence and dedication during revision and editing, to Gina Champagne, for her help in editing, and to Brian Bies, for his encouragement and guidance in formatting, outreach, and promotion.

Finally, I wish to thank the following people:

Mrs. Teresa Miles, who has inspired my love for taking risks and chasing creativity.

Mrs. Heather Farrington, who taught me that science, poetry, mathematics, and prose can dance to the beat of the determined student's heart.

Mrs. Mary Ellen Phelan, whose mentorship and development of my writing has allowed me to realize sincere appreciation and gratitude to be her student.

Professor Eric Koester, who coached me to "believe in bold."

Mr. Brian Brown, for being an inspiring teacher, supportive confidant, and compassionate friend. Without him, this book might still be found in the clouds of my imagination.

# APPENDIX

———

Aaronson, Scott. 6.845 Quantum Complexity Theory. Fall 2010. Massachusetts Institute of Technology: MIT OpenCourseWare, https://ocw.mit.edu. License: Creative Commons BY-NC-SA.

Barbastathis, George. 2.717J Optical Engineering. Spring 2002. Massachusetts Institute of Technology: MIT OpenCourseWare, https://ocw.mit.edu. License: Creative Commons BY-NC-SA.

Bartusiak, Marcia. Einstein's Unfinished Symphony. Yale University Press, 2017.

Bayne, Tim. The Unity of Consciousness. Oxford University Press, 2010.

Bertschinger, Edmund, and Hughes, Scott. 8.962 General Relativity. Spring 2006. Massachusetts Institute of Technology: MIT OpenCourseWare, https://ocw.mit.edu. License: Creative Commons BY-NC-SA.

Bertschinger, Edmund, and Taylor, Edwin. 8.224 Exploring Black Holes: General Relativity & Astrophysics. Spring 2003. Massachusetts Institute of Technology: MIT OpenCourseWare, https://ocw.mit.edu. License: Creative Commons BY-NC-SA.

Binzel, Richard. 12.400 The Solar System. Spring 2006. Massachusetts Institute of Technology: MIT OpenCourseWare, https://ocw.mit.edu. License: Creative Commons BY-NC-SA.

Blecher, Marvin. General Relativity: A First Examination. WSPC, 2016.

Böhmer, Christian G. Introduction to General Relativity and Cosmology. WSPC, Europe, 2016.

Butterworth, John. Smashing Physics: Inside the World's Biggest Experiment. Headline, 2014.

Carroll, Sean. A No-Nonsense Introduction to General Relativity. Enrico Fermi Institute and Department of Physics, The University of Chicago, 2001.

Chakrabarty, Deepto. 8.901 Astrophysics I. Spring 2006. Massachusetts Institute of Technology: MIT OpenCourseWare, https://ocw.mit.edu. License: Creative Commons BY-NC-SA.

Chen, Min. 8.811 Particle Physics II. Fall 2005. Massachusetts Institute of Technology: MIT OpenCourseWare, https://ocw.mit.edu. License: Creative Commons BY-NC-SA.

Cline, Jim. The Dark Energy of the Universe. January 2016. McGill University.

Du Sautoy, Marcus. The Great Unknown. Penguin Books, 2016.

Edelman, Gerald and Giulio Tononi. A Universe of Consciouness: How Matter Becomes Imagination. Basic Books, 2000.

Einstein, Albert. The Meaning of Relativity. Princeton University Press, 2004.

Etingof, Pavel. 18.238 Geometry and Quantum Field Theory. Fall 2002. Massachusetts Institute of Technology: MIT OpenCourseWare, https://ocw.mit.edu. License: Creative Commons BY-NC-SA.

Fang, Nicholas X. 2.71 Optics. Spring 2014. Massachusetts Institute of Technology: MIT OpenCourseWare, https://ocw.mit.edu. License: Creative Commons BY-NC-SA.

Feynman, Richard. QED: The Strange Theory of Light and Matter. Princeton University Press, 1988.

Feynman, Richard. The Feynman Lectures on Physics, Volumes I, II, III. Pearson P T R, 1970.

Field, Robert, and Tokmakoff, Andrei. 5.74 Introductory Quantum Mechanics II. Spring 2004. Massachusetts Institute of Technology: MIT OpenCourseWare, https://ocw.mit.edu. License: Creative Commons BY-NC-SA.

Fowler, Michael. PHYS 252, The Photoelectric Effect. University of Virginia: Department of Physics, http://galileo.phys.virginia.edu/classes/252/photoelectric_effect.html#References.

Gamow, George. Mr. Tompkins in Paperback. Cambridge University Press, 1965.

Guth, Alan. 8.286 The Early Universe. Fall 2013. Massachusetts Institute of Technology: MIT OpenCourseWare, https://ocw.mit.edu. License: Creative Commons BY-NC-SA.

Guth, Alan. 8.323 Relativistic Quantum Field Theory I. Spring 2008. Massachusetts Institute of Technology: MIT OpenCourseWare, https://ocw.mit.edu. License: Creative Commons BY-NC-SA.

Haller, George. 2.032 Dynamics. Fall 2004. Massachusetts Institute of Technology: MIT OpenCourseWare, https://ocw.mit.edu. License: Creative Commons BY-NC-SA.

Hanany, Amihay. 8.871 Selected Topics in Theoretical Particle Physics: Branes and Gauge Theory Dynamics. Fall 2004. Massachusetts Institute of Technology: MIT OpenCourseWare, https://ocw.mit.edu. License: Creative Commons BY-NC-SA.

Haus, Hermann, Melcher, James, Zahn, Markus, and Silva, Manuel. RES.6-001 Electromagnetic Fields and Energy. Spring 2008. Massachusetts Institute of Technology: MIT OpenCourseWare, https://ocw.mit.edu. License: Creative Commons BY-NC-SA.

Heisenberg, Werner. Physics and Beyond. Harper Torchbooks, 1971.

Hénault, François. "Simple alternative model of the dual nature of light and its Gedanken experiment," Proceedings of the SPIE, Volume 8121, ID. 81211J (2011). https://arxiv.org/pdf/1407.2522.pdf

Henderson, Linda Dalrymple. The Fourth Dimension and Non-Euclidean Geometry in Modern Art. MIT Press, 2013.

Hermundstad, Ann, Kinder, Jesse M., Nelson, Philip, and Bromberg, Sarina. From Photon to Neuron: Light, Imaging, and Vision. Princeton University Press, 2017.

Hu, Wayne. The Future of Cosmological Physics. January 2016. The Kavli Institute for Cosmological Physics at The University of Chicago.

Kaku, Michio. Hyperspace: A Scientific Odyssey Through Parallel Universes, Time Warps, and the 10th Dimension. Anchor Books, Knopf Doubleday Publishing Group, 1995.

Kaku, Michio. Physics of the Future: How Science Will Shape Human Destiny and Our Daily Lives by the Year 2100. Anchor Books, Knopf Doubleday Publishing Group, 2012.

Kong, Jin Au. 6.632 Electromagnetic Wave Theory. Spring 2003. Massachusetts Institute of Technology: MIT OpenCourseWare, https://ocw.mit.edu. License: Creative Commons BY-NC-SA.

Kong, Jin Au. 6.630 Electromagnetics. Fall 2006. Massachusetts Institute of Technology: MIT OpenCourseWare, https://ocw.mit.edu. License: Creative Commons BY-NC-SA.

Kong, Jin Au. 6.635 Advanced Electromagnetism. Spring 2003. Massachusetts Institute of Technology: MIT OpenCourseWare, https://ocw.mit.edu. License: Creative Commons BY-NC-SA.

Krauss, Lawrence. A Universe from Nothing: Why There is Something Rather Than Nothing. Free Press, 2012.

Lakatos, Imre. Proofs and Refutations: The Logic of Mathematical Discovery. Cambridge University Press, 1976.

Liu, Hong. 8.821 String Theory and Holographic Duality. Fall 2014. Massachusetts Institute of Technology: MIT OpenCourseWare, https://ocw.mit.edu. License: Creative Commons BY-NC-SA.

Livio, Mario. The Accelerating Universe: Infinite Expansion, the Cosmological Constant and the Beauty of the Cosmos. Wiley, 2000.

Mavalvala Nergis, and Greytak, Thomas. 8.03 Physics III. Spring 2003. Massachusetts Institute of Technology: MIT OpenCourseWare, https://ocw.mit.edu. License: Creative Commons BY-NC-SA.

McGreevy, John. 8.821 String Theory. Fall 2008. Massachusetts Institute of Technology: MIT OpenCourseWare, https://ocw.mit.edu. License: Creative Commons BY-NC-SA.

Nagal, Jennifer. Knowledge: A Very Short Introduction. Oxford University Press, 2014.

Norton, John D. HPS 0410 Einstein for Everyone: Special Theory of Relativity. February 2013. Department of History and Philosophy of Science, University of Pittsburgh.

Penrose, Roger. Cycles of Time: An Extraordinary New View of the Universe. Bodley Head, 2010.

Peterson, Ivars. Newton's Clock: Chaos in the Solar System. Freeman, 1993.

Polchinski, Joseph. String Theory. Cambridge University Press, 1998.

Rappaport, Saul. 8.282J Introduction to Astronomy. Spring 2006. Massachusetts Institute of Technology: MIT OpenCourseWare, https://ocw.mit.edu. License: Creative Commons BY-NC-SA.

Schechter, Paul. 8.284 Modern Astrophysics. Spring 2006. Massachusetts Institute of Technology: MIT OpenCourseWare, https://ocw.mit.edu. License: Creative Commons BY-NC-SA.

Sciolla. Gabriella. 8.022 Physics II: Electricity and Magnetism. Fall 2004. Massachusetts Institute of Technology: MIT OpenCourseWare, https://ocw.mit.edu. License: Creative Commons BY-NC-SA.

Shankar, Ramamurti. Fundamentals of Physics II. Yale University Press, 2016.

Shankar, Ramamurti. Principles of Quantum Mechanics. Plenum Press, 2011.

Silk, Joseph. The Infinite Cosmos: Questions from the Frontiers of Cosmology. Oxford University Press, 2006.

Smolin, Lee. Time Reborn: From the Crisis of Physics to the Future of the Universe. Allen Lane, 2013.

Stewart, Iain. 8.851 Strong Interactions: Effective Field Theories of QCD. Spring 2006. Massachusetts Institute of Technology: MIT OpenCourseWare, https://ocw.mit.edu. License: Creative Commons BY-NC-SA.

Stewart, Ian. In Pursuit of the Unknown. Basic Books, 2013. Strang, Gilbert. 18.06SC Linear Algebra. Fall 2011. Massachusetts Institute of Technology: MIT OpenCourseWare, https://ocw.mit.edu. License: Creative Commons BY-NC-SA.

Sudbery, Anthony. Quantum Mechanics and the Particles of Nature: An Outline for Mathematicians. Cambridge University Press, 1986.

Susskind, Leonard and Friedman, Art. The Theoretical Minimum: Quantum Mechanics. Basic Books, 2015.

Susskind, Leonard and Friedman, Art. The Theoretical Minimum: Special Relativity and Classical Field Theory. Basic Books, 2017.

Taleb, Nissam. The Black Swan: The Impact of the Highly Improbable. Allen Lane, 2007.

Tegmark, Max. 8.033 Relativity. Fall 2006. Massachusetts Institute of Technology: MIT OpenCourseWare, https://ocw.mit.edu. License: Creative Commons BY-NC-SA.

Tong, David. Lectures on String Theory. February 2012. The University of Cambridge.

Trefil, James. From Atoms to Quarks. Anchor Books, Knopf Doubleday Publishing Group, 1994.

Trout, Bernhardt. 10.675J Computational Quantum Mechanics of Molecular and Extended Systems. Fall 2004. Massachusetts Institute of Technology: MIT OpenCourseWare, https://ocw.mit.edu. License: Creative Commons BY-NC-SA.

Turner, Michael. Science with a Wide-Field Infrared Telescope in Space. Undated. The Kavli Institute for Cosmological Physics at The University of Chicago.

Wang, Wei-Chih. ME557 Electromagnetic Wave Theory. University of Washington: Department of Mechanical Engineering, https://depts.washington.edu/mictech.

Williamson, Timothy. Knowledge and Its Limits. Oxford University Press, 2000.

Woit, Peter. Not Even Wrong: The Failure of String Theory and the Continuing Challenges to Unify the Laws of Physics. Jonathan Cape, 2006.

Zee, Anthony. Fearful Symmetry. Princeton University Press, 2017.

Zwiebach, Barton. 8.04 Quantum Physics I. Spring 2016. Massachusetts Institute of Technology: MIT OpenCourseWare, https://ocw.mit.edu. License: Creative Commons BY-NC-SA.

Zwiebach, Barton. 8.05 Quantum Physics II. Fall 2013. Massachusetts Institute of Technology: MIT OpenCourseWare, https://ocw.mit.edu. License: Creative Commons BY-NC-SA.

Made in the USA
San Bernardino, CA
25 February 2019